HANDBOOK OF STORAGE TANK SYSTEMS

HANDBOOK OF STORAGE TANK SYSTEMS

Codes, Regulations, and Designs

edited by

Wayne B. Geyer

Steel Tank Institute
Lake Zurich, Illinois

Associate Editor

Jim Wisuri

Prairie Oak Communications, Inc.
River Forest, Illinois

Sponsored by
Steel Tank Institute
Lake Zurich, Illinois

MARCEL DEKKER, INC. NEW YORK • BASEL

ISBN: 0-8247-8589-4

This book is printed on acid-free paper.

Headquarters
Marcel Dekker, Inc.
270 Madison Avenue, New York, NY 10016
tel: 212-696-9000; fax: 212-685-4540

Eastern Hemisphere Distribution
Marcel Dekker AG
Hutgasse 4, Postfach 812, CH-4001 Basel, Switzerland
tel: 41-61-261-8482; fax: 41-61-261-8896

World Wide Web
http://www.dekker.com

The publisher offers discounts on this book when ordered in bulk quantities. For more information, write to Special Sales/Professional Marketing at the headquarters address above.

Current printing (last digit):
10 9 8 7 6 5 4 3 2 1

PRINTED IN THE UNITED STATES OF AMERICA

Preface

The storage of hazardous liquids has never been more reliable. This assertion reflects the tremendous progress over the last two decades in developing tanks and appurtenances (in response to regulations and standards) that incorporate the best thinking that leads to safe, long-term performance. However, that does not mean there will never be another release from an underground or aboveground storage tank system. New products and services will continue to surface as the industry defines even higher standards of performance.

The *Handbook of Storage Tank Systems* reflects the invaluable contributions of experts in standards, manufacturing, installation, and specification of storage tank systems. Each author deserves our thanks for shedding light on the best equipment and methods for storing or handling petroleum or chemicals.

Behind the scenes, a number of organizations and individuals merit recognition for enabling this book's publication. The American Iron and Steel Institute and the Steel Tank Institute helped to fund the project. Jeanne McFadden and Ann Pulido served as the production editors who guided the book to completion. Significant out-of-the-limelight contributions were provided by Arlene Barnhart, Athena Bolton, Charles Frey, Jr., John Hartmann, Julie Hoffmann, Diane Lekovish, Vicky Lekovish, Jack Quigley, Bob Renkes, Dana Schmidt, Rick Sharpe, and Lorraine Waller.

A special note of appreciation goes to Wayne Stellmach, who, in addition to authoring two important chapters, provided invaluable contributions in developing the overall content, making arrangements with authors, and managing the book's progress.

Jim Wisuri

Disclaimer

Every effort has been made by the contributors and the editors to ensure the accuracy and reliability of the information in this book. However, neither the authors, the editors, the Steel Tank Institute (STI), nor Marcel Dekker, Inc., make any representation, warranty, or guarantee in connection with the information herein. The authors, editors, STI, and Marcel Dekker, Inc., hereby expressly disclaim any liability or responsibility for:

Loss or damage resulting from the use of information in this book
The violation of any federal, state, or local regulation with which the information in this book may conflict
The infringement of any patent resulting from information in this book

The rules and regulations governing storage tank systems vary widely from one jurisdiction to the next, and they change—sometimes from one year to the next. In some jurisdictions, a fire-code inspector may require the owner of a storage tank system to address issues that seemingly contradict the mandates of a building-code official or an environmental regulator. The most prudent course of action, in all cases, is to check with regulators first before finalizing a specification, purchasing storage-system equipment, or installing tanks, piping, and other appurtenances.

In addition, it is recommended that all applicable standards be checked for details that may affect an individual project. For best results, development of tank-systems design and installation should rely on the assistance of professionals who have proven experience with storage tank and liquid handling systems.

Contents

Preface *iii*
Contributors *ix*

SHOP-FABRICATED TANKS

1. Introduction: A History of Storage Tank Systems 1
Wayne B. Geyer

2. Historical Perspective on Standards and Codes 21
Marshall A. Klein

3. History of the Uniform Fire Code 29
Lavern Cary

4. Quality Control on USTs and ASTs 33
Larry O'Shea

5. UST and AST Fabrication 41
Randy Smith and Larry O'Shea

6. Storage Tank Specification Considerations 49
Wayne Stellmach

7. To Bury or Not to Bury: Steel Tank Technology Decisions 91
Wayne B. Geyer

UNDERGROUND STORAGE TANKS

8. UST Design: The Wall Thickness Question 99
Lorri Grainawi

9. Development of UL Standards for Underground Steel Tank Safety 105
Shari Hunter and John J. Hawley

10. Development of ULC Standards for Underground Storage and Handling of Flammable and Combustible Liquids 113
Gordana Nikolic

11. Federal UST Regulatory Program 121
Marcel Moreau

12. Steel UST Technologies 137
Jim Wisuri

13. Underground Storage Tank Installation 143
Jim O'Day

14. UST Performance 153
Brian C. Donovan

15. Corrosion and Cathodic Protection on Underground Storage Tanks 161
Taylor Leon

16. How to Specify UST System Equipment 175
Alex Ralston

ABOVEGROUND STORAGE TANKS

17. AST Design 183
Lorri Grainawi

18. Development of UL Standards for Aboveground Steel Tank Safety 193
Shari Hunter and John J. Hawley

19. Development of ULC Standards for Aboveground Storage and Handling of Flammable and Combustible Liquids 199
Gordana Nikolic

20. Secondary Containment for Noninsulated Steel Aboveground Storage Tank Systems 209
Jerry L. Waller

21. AST Environmental Regulations 217
Wayne B. Geyer

22. Plan Review and Inspection of Aboveground Storage Tanks 225
Scott A. Stookey

23. Aboveground Storage Tank Installation 293
George H. Watkins

24. How to Specify AST System Equipment 311
Charlie Glab

Appendix 327
Wayne Stellmach

Index *339*

Contributors

Lavern Cary Cary & Associates, Sherwood, Oregon

Brian C. Donovan Containment Advisors, Inc., Lake Zurich, Illinois

Wayne B. Geyer Steel Tank Institute, Lake Zurich, Illinois

Charlie Glab Morrison Brothers Company, Dubuque, Iowa

Lorri Grainawi Steel Tank Institute, Lake Zurich, Illinois

John J. Hawley Engineering Services, Underwriters Laboratories, Inc., Northbrook, Illinois

Shari Hunter Certification Services, Underwriters Laboratories, Inc., Santa Clara, California

Marshall A. Klein Marshall A. Klein & Associates, Eldersburg, Maryland

Taylor Leon Steel Tank Institute, Lake Zurich, Illinois

Marcel Moreau Marcel Moreau Associates, Portland, Maine

Gordana Nikolic Underwriters' Laboratories of Canada, Scarborough, Ontario, Canada

Jim O'Day O'Day Equipment, Inc., Fargo, North Dakota

Larry O'Shea Steel Tank Institute, Lake Zurich, Illinois

Alex Ralston Petcon, Inc., Jackson, Mississippi

Randy Smith Hamilton Welding Company, Columbus, Ohio

Wayne Stellmach Rolling Meadows, Illinois

Scott B. Stookey Hazardous Materials Section, City of Austin Fire Department, Austin, Texas

Jerry L. Waller Modern Welding Company, Inc., Owensboro, Kentucky

George H. Watkins The Watkins Company, Fayetteville, Georgia

Jim Wisuri Prairie Oak Communications, Inc., River Forest, Illinois

1

Introduction: A History of Storage Tank Systems

Wayne B. Geyer
Steel Tank Institute, Lake Zurich, Illinois

I. BACKGROUND

In August 1859, the first oil well was constructed in Titusville, PA. The visionaries who financed and developed the primitive derrick and drill believed that "rock oil" would provide an excellent source of energy for illuminating buildings [1]. And for a few years, it did—until Thomas Edison found a way during the early 1880s to harness electricity. Luckily for the oil industry, other world-changing inventors in Europe and the United States had already begun the first steps toward redefining transportation—through development of a four-stroke engine and adaptation of the motor to power a buggy. The seeds of an automotive industry had been sown. And, from that point on, a need for storing petroleum products grew.

The first service stations required minimal tank storage capacity. In fact, it was common for product to be stored within the dispenser itself. As the need for hydrocarbons grew, the ability to store the product safely became an important growth factor for the petroleum and automotive industries. The storage tank industry traces its start to these events that have altered society.

II. FORMATION OF IMPORTANT ASSOCIATIONS AND STANDARDS

In 1916, a group of Midwestern tank and boiler fabricators established a trade group, known as the National Steel Tank Association. Today this group is known as the Steel Tank Institute (STI), an international trade association enhancing the value of steel storage tank fabricators and suppliers of products and services to the industry. The Institute serves members, the tank industry, regulators, and the public through promotion of state-of-the-art products, standards, education, quality certification, legislative initiatives, and promotion of steel as a viable material for storage of products today and tomorrow.

Most of the product storage applications discussed within this book pertain to hydrocarbons, chemicals, and other flammable and combustible liquids. Most of these tanks are vented to the atmosphere. Tanks are also produced for various processes, such as separation of oil from water. Of course, tanks can also be built to withstand high pressures.

As a general rule of thumb, tanks 50,000 gal and under can be fabricated at a manufacturing facility and shipped to the job site in one piece on a truck. Highways usually limit the tank diameter to 12 ft. A typical tank size installed at a retail service station is an 8-ft diameter by 32-ft in length for a 12,000-gal tank. Tanks with capacities greater than 50,000 gal are erected in the field—and are beyond the scope of this book. Most field-erected tanks store either water or oil.

Shortly after the time of STI's formation, a third-party testing laboratory was developing its first safety standards for atmospheric steel storage tanks. Underwriters Laboratories (UL) is a testing agency specializing in certifying product safety through a listing procedure. In 1925, the agency published the first edition of its UL 58 document (Standard for Steel Underground Tanks for Flammable and Combustible Liquids). A few years earlier, in 1922, UL developed its first atmospheric aboveground steel storage tank standard, UL 142 (Steel Aboveground Tanks for Flammable and Combustible Liquids).

One other standard critical to the development of the storage tank industry was created at the start of the 20th century by a group that developed and promoted industry consensus standards for protection of the public and firefighters from the hazards caused by fires. The National Board of Fire Underwriters, formed in 1866, developed NBFU 30 (circa 1904), which, as its title states, covered "Rules and Requirements for the Construction and Installation of Systems for Storing 250 Gallons or Less of Fluids Which at Ordinary Temperatures Give Off Inflammable Vapors, as Recommended by its Committee of Consulting Engineers" [2]. These standards were developed in the early 1900s, but the most significant technological changes for both underground and aboveground storage tanks did not take place until more than half a century later.

Codes and standards developed by NBFU over time became the responsibility of the National Fire Protection Association (NFPA), which modified the storage system standard's code designation to NFPA 30L (published in 1913). Today that standard is known as NFPA 30 (Flammable and Combustible Liquids Code), which was first published in 1957. Many governmental bodies today reference the NFPA standard either by legislation, statutes, or ordinances to give local officials an enforceable code document. The code refers to various construction standards written by UL and STI as acceptable construction practices. NFPA's work on storage tank issues has been essential, as most hydrocarbons are either flammable or combustible, requiring safe storage, handling, and management.

As the use of petroleum products escalated, the producers of hydrocarbons developed their own association. The American Petroleum Institute (API) was formed in 1919. Today API is a huge association representing major oil companies worldwide. API members are interested in refining, transportation, storage, and the use of fuels for motor vehicles, heating and other related uses. API has also published a number of standards pertaining to the safe storage and handling of flammable and combustible liquids.

The Petroleum Equipment Institute was formed in 1951. PEI represents petroleum equipment distributors and manufacturers throughout the world. Many of PEI's members also install petroleum equipment. After 1985, PEI developed several important guideline documents considered essential for the proper installation of underground and aboveground storage facilities.

Of course, there are many other associations with close ties to the industry. And a number of other fire and mechanical codes also exist that will be mentioned later in this handbook.

III. STEEL TANK CONSTRUCTION PRIOR TO WORLD WAR II

In the "early days" of steel tank production, the joining of steel by riveting was common. Most tanks were small by today's standards, under 1100 gal, but a few larger tanks were built (Fig. 1). Galvanized steel sheet was often used in producing the tank. However, World War II created a shortage of galvanized product. UL began to allow the manufacture of tanks from black carbon steel owing to the scarcity of galvanized steel.

IV. STEEL TANK CONSTRUCTION AFTER WORLD WAR II

The innovation of heating homes with natural gas was becoming popular at this same time. Yet, the majority of homes outside dense urban areas relied upon heating oil, which often was stored in the basement of the building or residence. In 1927, Underwriters Laboratories had developed its UL 80 standard for basement heating oil tanks that were usually obround in shape and under 660 gal in capacity. But the tanks tended to have a shorter life than desired because of internal corrosion. Water would accumulate in the bottoms, and little or no maintenance on the tank or the removal of water would take place.

Figure 1 Riveted tank transported by horse carriage in the early 1900s.

In the mid-1950s, STI advocated placing the product line on the bottom of the tank to prevent any accumulation of water. This design breakthrough carried the water with the heating oil to the burner, which led to evaporation. And in 1956, STI issued its Midwest 56 Standard, a widely recognized approach to designing the size and location of openings along the top of an underground storage tank. These efforts were the earliest recorded ventures by the Institute in impacting standards. Some fuel tanks were installed aboveground at gas stations, but many were installed below ground, for aesthetic purposes, particularly at public retail service stations.

Tanks installed in the 1950s were generally coated with red lead primer or a thin asphaltum-based paint. These coatings were fine in preventing atmospheric corrosion; however, they were nearly useless for protection against corrosion in many underground environments. Unfortunately, some tank buyers continued to install non-protected tanks underground until Congress in 1984 empowered the U.S. Environmental Protection Agency to prohibit such tanks.

There were some early attempts to protect underground tanks from corrosion in the 1950s. One PEI member developed a special kit of galvanic magnesium anodes designed for cathodic protection of the tank via connection in the field to a steel tank or steel pipe. However, there were few takers in that era, and the business eventually pursued other profit-making activities.

In the early 1960s, the average atmospheric tank size had increased to nearly 4000 gal, and a new material was being used to develop and test a different design of an underground flammable and combustible liquid tank. The tank was nonmetallic. It was made from fiber-reinforced plastic (FRP). Tank buyers hoped to avoid the inevitable problem with steel underground storage tanks releasing product due to corrosion. At that time, environmental concerns did not drive the new product's development. Foremost on the mind of petroleum marketers were conservation and the cost to replace lost product.

V. CORROSION CONTROL OF UNDERGROUND STEEL STORAGE TANKS

It took several years to make the FRP tank design acceptable enough for use at service stations. The first prototypes lacked structural integrity. Reinforcing ribs were later added to the design to enhance the tank's stiffness. By the latter half of the 1960s, UL had issued its first listing of a nonmetallic underground storage tank for flammable and combustible liquids. Within a year or two, NFPA 30 accepted the nonmetallic tank as an alternative. Up until that time, all tanks had to be steel, usually built to the UL 58 standard. More than a decade later, UL published the first edition of a new standard, UL 1316—(Glass-Fiber-Reinforced Plastic Underground Storage Tank for Petroleum Products).

The steel industry paid close attention to the efforts to produce a nonmetallic tank, as did major oil producers. So steel tank makers responded with their own research efforts—focused on methods to make steel tanks corrosion resistant.

One effort was to install the steel tank in a plastic wrap or baggie. One major oil company had installed a number of these types of systems in Michigan as early as 1959. Steel tanks from this initiative were uncovered about a dozen years later and found to be free of any corrosion. Preventing groundwater and corrosive-soil contact against the outer steel tank surface essentially eliminates one of the necessary ingredients for corrosion to take place.

By the early 1970s, the plastic baggie concept was commercialized for nationwide steel tank production. More than 1000 plastic-wrap tanks had been installed by 1970. Although this concept had merit, the plastic was quite thin and prone to tearing. Pieces of plastic were often "sealed" together with tape. Of course, if one did not keep the steel surface completely free from contact with moisture or soil, a guarantee that the tank would be corrosion-resistant was virtually meaningless.

The US Steel Company had a major interest in seeing that steel was the material of choice for storage applications. Faced with the nonmetallic threat, US Steel dedicated personnel from its research center to corrosion control for underground storage tanks. One US Steel concept was to bury the tank in a high-pH limestone backfill. This approach depended upon nature taking its course—enabling the limestone to react with the steel and form a calcareous layer or coating that would protect the tank from corrosion. This was a great idea, except when one began to examine how tanks were installed. If a clod of native clay or trash were to be inadvertently mixed in the limestone and in contact with the steel, a corrosion cell would form. The surface area of the tank in contact with the clay would become anodic and corrode at a more rapid pace than if the tank were entirely buried in clay. With the number of tanks installed by contractors with no formal training in corrosion control, the risks of such a technology were too great.

Some fabricators investigated the concept of cladding the steel with another metal, such as stainless steel. But this concept would have been extremely expensive. Underground tanks were considered a commodity item during that era.

STI developed a standard for coating a steel tank with a thick fiber-reinforced plastic coating. The concept was simple. If a nonmetallic tank could be installed to eliminate corrosion, why not coat the steel tank with that very same material? By eliminating soil or groundwater contact with the external steel tank, corrosion of the surface cannot take place. STI wrote the standard in 1968 and called it STI-LIFE. The standard actively existed for about five years. But fabricators stopped using it for the most part because of aesthetics and cost. The glass strands were not always lying flat and parallel to the steel. Equipment to produce the coating was relatively unsophisticated, unlike today's manufacturing processes in which the steel tank rotates while glass fibers and resin are sprayed onto the exterior.

Several fabricators continued to make FRP-coated steel tanks. Since the tank did not use anodes, tank buyers had no corrosion-control maintenance requirements. With the advent of the U.S. Environmental Protection Agency's rules in 1988, which required testing and record keeping for cathodically protected tanks, the FRP-coated steel tank became a popular product in the 1980s. A different standard was developed in 1987 to represent this type of composite product, which became known as the ACT-100 tank.

However, the STI-LIFE tank prompted STI to develop a membership technology initiative, which set the stage for many of the Institute's current tank-technology programs. The STI-LIFE effort included tank production specifications, STI labels and certification numbering, STI random quality control inspections and noncompliance penalties, and licensing and trademarks.

But the steel tank industry's greatest success followed the development of a cathodically protected underground storage tank. In 1969, the Steel Tank Institute developed its sti-P_3® standard. The design consisted of three major elements: good dielectric coating, galvanic anodes, and electrical isolation of the tank from steel piping. The earliest coating was a coal-tar epoxy, which was significantly advanced from the industry's prior products of choice [3]. Dow Chemical and US Steel were involved in the development of the sti-P_3 design with Kennedy Tank, an STI member. Dow Chemical brought its expertise in magnesium anodes to the table. The system was designed to withstand 5 percent coating damage—which is very high by today's standards. Because of innovations in coating durability—and greater professionalism among installers—most tanks go in the ground today with no damage and no steel surface directly exposed to the soil.

There were a number of improvements in the P_3 design during the 1970s and 1980s. The Steel Tank Institute developed a national registration and marketing program for the nine nationwide companies that signed up for the technology in 1969. STI provided tank owners with a 20-year national warranty against failure due to external corrosion. With the warranty came a strong quality assurance inspection program. The fabricator had to fill out a special inspection form for each tank produced. Eventually, STI hired UL to perform its inspections—until early in the 1980s, when STI started its own inspection program. Inspectors arrive at a tank shop unannounced and at random to verify that the fabricator is meeting STI's requirements. Today, the industry norm for the warranty period of underground storage tank products is 30 years.

STI developed a coating test procedure to allow application of other coating types to the P_3 tanks. Shortly after 1980, a fast-cure urethane coating technology was approved for use with sti-P_3 tanks. This coating revolutionized steel tank production. Due to its immediate curing ability, it enabled a tank to be shipped the same day it was coated, an unprecedented distribution breakthrough. Also, STI developed a better surface preparation specification to assure proper adhesion of the coating to the tank, and therefore, a longer tank life.

In 1983, STI adopted zinc weld-on anodes for the P_3 tank. Until that time, all anodes were wire connected to the tank and made from magnesium. The weld-on concept offered some dramatic cost savings through attaching the anode at the factory. But it also provided a much better means for the anode to remain connected to the tank during installation. Some installers simply did not know what the wired-on bag anode was. In some cases, they actually disconnected the tank from the anode. The welded-on anode eliminated this misunderstanding.

Throughout the late 1970s, more and more media attention was focused on leaking underground storage tanks—heightening the tank industry's emphasis on environmental protection. Many efforts were directed toward integrity testing of the tank. But tank integrity testing was fairly new and had limited accuracy. Large leaks were easily noticed, but microscopic leaks were difficult to spot. This resulted in record-keeping improvements for monitoring inventory. Many of the major oil companies began replacing their bare steel tanks with protected tank technology—cathodically protected steel and nonmetallic FRP. In some cases, existing tanks were retrofitted with an internal lining, or a field-installed, impressed-current cathodic protection system. However, no national standards existed for either approach.

By 1977, many members of the industry had become aware of leaking underground storage tanks. Fire codes, such as NFPA 30 and the Uniform Fire Code, introduced language into their respective codes mandating corrosion protection. But by current standards, fire codes in the late 1970s were lax. For instance, the codes said steel tanks could be cathodically protected, simply well coated, or not protected from corrosion at all—each of the three instances dictated by the soil corrosivity for the site at which the tank would be installed [4]. By this time, NFPA had also produced a standard, NFPA 329, for investigating and locating releases from flammable and combustible liquid storage tanks. This recommended practice also covered tank integrity precision testing—the only document of its kind at that time.

Many local, state, and county officials by 1982 were involved with leaking underground storage tanks through enactment of regulations. In Florida and California, several areas were starting to develop requirements for secondary containment, which was still in its infancy.

However, some steel tanks had been made with secondary containment during the early 1970s. In addition, some tanks were insulated with an outside weathering jacket placed over the insulation to form an additional barrier. But mass production of dual-walled steel tanks to prevent releases into the environment was rare.

In 1983, Steel Tank Institute formed a special strategic planning committee to address the new environmentally driven concerns with underground storage tanks. One committee was formed to develop recommendations for the corrosion protection of steel piping. A second committee was formed to reexamine the FRP-coated steel composite tank. But the most important committee was formed to address secondary containment.

At that time, most designs being developed provided 110 percent containment of the primary tank. This required the tank to have channel or angle iron separating the primary and secondary tank. Each tank was independent—capable structurally of standing by itself. STI discovered that German tank makers were already building to a double-wall design. STI visited and learned from the German experience, which led to America's first secondary containment standard, which was published in 1984.

STI's standard is relatively simple. The outer wall of steel is intimately wrapped over the primary tank and can be of a thinner gage. Due to the intimate wrap, the two walls act as a single structural unit, reducing the costs to build the tank. By adding a monitoring port, the owner can detect leaks in a number of ways, either manually or with special continuous monitoring equipment [5]. This enabled new technology to evolve. Electronic and mechanical means to detect liquids followed. STI verified that the intimate wrap did not inhibit product or water introduced into the interstice from finding its way to the monitoring port. Hence, to gain acceptance, monitoring equipment had to be able to detect liquids in the interstice automatically, although such detection could occur easily by manually inserting a gage stick into the monitoring port with a dry cloth tied to the bottom.

The 1980s were a time of unprecedented change in the industry as new standards and code language redefined the best practices for operating and installing UST systems. By 1984, the NFPA and UFC had modified their codes to require protection of both tanks and piping from corrosion, without the exceptions and dependency of soil resistivity. By 1985, the National Association of Corrosion Engineers (NACE) developed their first standard for corrosion control of steel underground storage tanks. NACE RP 0285 was written very generically to allow all types of factory- and field-engineered cathodic protection systems to be installed by qualified corrosion engineers (6). NACE had earlier developed a standard for underground piping. The document, RP 0169, was developed as a response to U.S. Department of Transportation requirements for hazardous liquids and gases within pipe systems, i.e., cross-country natural gas, oil, and gasoline transmission lines, etc.

VI. LEGISLATION AND REGULATION OF USTS

In 1984, the Congress enacted law to regulate underground storage tanks. By that time, media attention was reaching a peak. The CBS-TV show "60 Minutes" had aired an exposé of several situations of leaking USTs and their expensive impact on local communities. Congress gave the EPA authority to develop regulations.

In 1985, EPA released the Interim Prohibition that required newly installed tanks to be corrosion resistant, compatible with the product stored and structurally sound. It provided guidance while the EPA developed technical requirements to meet the legislative initiative. By that time, several states had already promulgated their own regulations for USTs. For example, Florida and California had already mandated secondary containment systems. Kansas had already required tank-overfill prevention. Connecticut and New York developed their own requirements for USTs. As a matter of fact, New York was one of the first states to publish a comprehensive technical manual on USTs.

A number of other environment protection options for UST systems existed in the mid-'80s, but are seldom used today:

1. Tank tightness testing. Some of the most prominent integrity tests required tanks to be filled to a level above the tank top as fluid level and temperature would be closely monitored over several hours. Today, many tank tests are performed with the tank liquid level far below its capacity and with a slight vacuum drawn.
2. Excavation liners. Tank excavation membrane liners were an option to provide secondary containment for tanks and pipe. Today, the tank and its piping are typically addressed as individual tank system components employing integral secondary containment.
3. Overfill prevention. Tank overfill prevention was handled through a ball valve located under the vent pipe connection at the top of the tank. As product came in, the ball would eventually rise into the opening and cause a restriction in the flow rate coming into the fill. Today, most systems use an overfill prevention valve within the tank and in conjunction with a spill container.
4. Leak detection. Release detection from tank and piping was normally accomplished through manual inventory and perhaps a pipe leak detector attached to the submersible turbine pump. Although the pipe leak detector is still commonly used, inventory methods have become quite sophisticated. Inventory levels are electronically tracked. New statistical inventory reconciliation methods are used to analyze inventory records and determine any unusual releases, theft or malfunctioning metering equipment. Another method of tank or piping release detection that has fallen out of favor was the use of monitoring wells to observe groundwater or sniff for vapors.

EPA commissioned a number of studies on the causes of underground storage tank releases. Most notable was the Summary of State Reports on Releases from Underground Storage Tanks. This 50-state study indicated that corrosion and structural failure were the two greatest causes of releases. However, a couple of other findings began to take hold as principal causes of "UST" releases: one, that releases from piping (improperly installed or corroded) were substantial, and two, that petroleum product had entered soil, local aquifers, and surface waters from tank overfill releases [7]. The study cited structural failure as the most common reason for problems with nonmetallic tanks. It was noted that such failures were catastrophic—often leading to the release of a tank's entire contents into the environment. The study also provided details on the life of steel storage tank systems. The average non-corrosion-protected tank would last up to 15 years before it failed. Of course, the life of the tank was quite dependent on the steel tank thickness (dictated by size), the soil environment in which the tank was placed, and the installation methods.

Two years later, a study from Suffolk County, New York, added to the growing body of knowledge on tank life. The Long Island study of 500 USTs removed from service showed that small heating-oil tanks, employing lighter-gage steel, were more likely to perforate than tanks with larger capacity. On the other hand, the Suffolk County study indicated that some unprotected steel tanks over 50 years of age were removed without having failed [8]. Similar studies were performed in Ohio and Pennsylvania with similar findings [9,10].

These studies were useful tools for EPA to determine regulatory program requirements. Their findings paved the way in 1987 for the agency to issue a draft of the federal regulations. Key regulatory provisions emphasized corrosion protection for tanks and piping, structural integrity, release detection, proper installation, and corrective action. EPA also addressed existing tanks and the time frames by which tank owners had to comply. Though the regulations generally allowed tank owners 10 years to comply with the new rules (such as the corrosion protection mandates), some provisions required earlier action for existing tank systems. For example, release detection of pressurized pipe systems had to come into compliance with testing and release detection methods several years after rule publication. EPA officials believed pressurized systems compelled earlier deadlines because releases from such piping were substantial. The pumping of liquid into the environment, which was difficult to detect via tank inventory control records, required an advanced deadline because the pressurized approach created a significantly greater environmental-risk factor.

EPA referenced several national standards including STI's sti-P_3 tank specification and STI's dual-wall tank requirements. EPA also referenced several UL tank construction standards (UL 58, UL 1746, and UL 1316) and recommended practices for installation from API and PEI. In addition, EPA referenced fire codes that included existing requirements for USTs.

EPA also added financial responsibility requirements. Tank owners had to have $1 million in financial wherewithal to meet the regulation. This came at a time when pollution and product liability insurance coverage for UST systems was hard to come by. As a result, many states passed legislation enabling the formation of government-administered trust funds designed to help pay for expensive cleanups, which by 1998 were averaging more than $130,000 per site.

EPA published its draft of regulations and gave all interested parties about one year to comment on the proposal. EPA hosted three public hearings on the regulation. In some cases, the agency made changes to address unforeseen problems and needs. The rule went into effect on December 22, 1988 [11]. The new regulations referenced several tank-manufacturing standards that did not exist before, and did not appear in, the 1987 draft. EPA listed three generic corrosion control designs for USTs—FRP, cathodic protection, and composite. With each reference came a list of national tank standards that would meet the EPA requirements.

For composite tanks, the ACT-100 standard was referenced. The ACT-100 tank was developed in 1987 by a new trade association called the Association for Composite Tanks. In November 1989, UL published its first edition of UL 1746 (External Corrosion Protection Systems for Steel Underground Storage Tanks). This standard covered corrosion prevention via cathodic protection, composite tanks and jacketed secondary containment technology.

Another new standard focused upon the internal lining of underground tanks. Several tank-lining companies formed the National Leak Prevention Association and created the NLPA 631 standard. A reference to an API standard for tank linings was also made. Several important features of the NLPA standard included minimum requirements for the condition of the tank to enable installation of a lining. If corrosion holes were extensive in size or number, the tank could not be lined. Safety in entering a tank was also a very important issue.

EPA also allowed a means for new technology to become accepted, which was very important because new approaches to UST systems were surfacing rapidly. This coincided with the emergence of smaller, faster and more powerful computers that provided new automated solutions, especially in leak detection. The tank and pipe industry also witnessed an influx of new technological approaches, such as jacketed steel tanks. A jacketed tank is an outer, nonmetallic containment for a steel tank that also provides corrosion protection. The most notable materials developed for secondary containment were high-density polyethylene and fiber-reinforced plastic. Typically, these systems were shipped from the factory with a vacuum between the walls of the tank. The most notable jacketed tank systems had trade names such as Total Containment, Permatank, Glasteel II, and Elutron. The latter three were FRP-jacketed steel. The jacketed steel tank offered several advantages—lighter weight than a double-wall steel tank and a lower cost. Steel Tank Institute developed a national specification for the Permatank system.

With each technology came associated patents to protect the invention. Days of old were gone. Tanks had become big business. UL 1746 became critical to the development of jacketed tanks. Authorities having jurisdiction looked for the UL label to get assurance that the tank met third-party national test standards.

Pipe systems also experienced a major market change. With the introduction of EPA's regulations, galvanized steel pipe, the predominant materials used in the 1970s and the first half of the 1980s became a dinosaur, even though entities such as STI and PEI had developed recommended practices to provide guidance on cathodic protection of steel coated pipe systems. As the new regulations took effect, FRP pipe took over the market. However, the new technology boom continued. By the early 1990s, new flexible pipe designs were introduced. Flexible pipe was made of various elastomers and plastics, and shipped to a site in coils. Its advantage was the elimination of buried field joints, a continuing troublesome area for system releases. By the mid-1990s, several flexible pipe systems were available and capturing a considerable share of the UST system piping market. One other notable construction feature emerged as the EPA regulations gained prominence. Sumps and boxes were placed above the tank and under the dispenser to catch releases from fittings and maintenance activities.

Steel Tank Institute in 1986 was the first organization to develop a national sump design standard, known simply enough as STI-86. It was designed to allow all fittings and important tank appurtenances to be clustered in one spot with the protection of secondary containment. This included the submersible turbine pump, vapor recovery equipment, gages, and fill openings. The sump container was made from steel, and would "catch" any releases from the enclosed equipment. In addition, secondary containment pipe would terminate in the sump where sensors were mounted to detect releases from piping as well [12]. The STI-86 paved the way for a wave of technological innovation. Within a few years, the STI-86 became obsolete—replaced by lightweight plastic sump containers.

By the time 1990 rolled around, new technology and new trends with storage tank systems were commonplace. But the final decade of the 20th century unexpectedly unveiled a completely new trend—aboveground storage tanks. Because of the negative headlines associated with expensive UST cleanups—including contaminated soils and water resources—tank owners began to think seriously about the advantages and disadvantages of owning an underground tank system. Many elected to close their tanks and, in the case of fueling vehicles, drive to the nearest public retail service station.

VII. REGULATION OF ASTS

Other tank-system owners began to study the advantages of an aboveground tank system. There were several characteristics that made an AST alluring. First, the owner/operator could see all surfaces of the tank. There was no need to depend on other equipment or contractors to verify a tank system was free from releases. Visual release detection is simple, cheap, convenient, and trustworthy. Secondly, the owner/operator did not have to meet the financial responsibility requirements of regulated UST systems. And third, the owner/operator didn't have to worry about expensive soil cleanups, which seemed to be constantly in the public eye via local and national headlines. Finally, many owner/operators felt that aboveground tank systems were cheaper to install and less regulated. The new era of aboveground tanks was about to begin.

On the other hand, many local jurisdictions did not allow aboveground fueling systems. They followed various national fire codes that carried either severe restrictions upon ASTs, or simply did not allow the aboveground storage of fuel. Despite that, fleet owners and small aviation fueling facilities saw tremendous benefit with owning ASTs, as described above. The typical owner of a public-accessible retail service station continued to prefer underground storage tanks (as did most fire inspectors) over potentially unsightly and potentially unsafe aboveground tanks.

As the compliance difficulties of UST owners became evident, some states began to express sympathy through legislation to ease restrictions upon the use of aboveground tanks. Thus, lawmakers in several states completely bypassed the fire prevention safeguards built into local and national fire codes.

In some cases, such as the Uniform Fire Code, no fueling of motor vehicles from ASTs was allowed. Others, such as NFPA, did provide some exceptions for tanks with capacity smaller than 6000 gal and often used in rural applications at commercial, industrial, and governmental facilities (also known as private fueling systems). With the suddenly strong demand for aboveground tanks, the codes needed to find a way to allow the safe siting of aboveground fueling facilities at service stations.

Major additions were added to codes in the early 1990s. NFPA saw the need as so urgent that a Tentative Interim Amendment, or TIA, was issued in 1992 to allow an aboveground tank to be installed inside a concrete room, whether that room was located above or below grade. By 1993, NFPA had added code language to allow other tanks to be installed aboveground, including traditional UL 142 tanks, and another new technology, fire-resistant tanks.

At the same time, the Uniform Fire Code was also modifying provisions that would lead to increased usage of aboveground tank storage. The UFC previously did not allow any AST fueling facilities, except special enclosures inside buildings. A special enclosure was defined as a 6-in.-thick concrete enclosure over a steel tank; a fairly common application was in parking garages. The logic was that if a special enclosure was acceptable inside a building, why wouldn't it be acceptable outside?

The UFC developed a special code appendix to accommodate ASTs. The goal was to emulate the fire safety obtained from an underground storage tank, which was completely backfilled and free of risks from vehicles, vandals, and fires. An associated fire test procedure was developed to fulfill the safety needs. Tanks had to employ secondary containment and insulation [13].

This was the birth of "protected tanks." Several Uniform Fire Code criteria defined a protected AST: (1) It had to prevent an internal tank temperature increase of more than 260°F when the structure was exposed to a 2000°F two hour fire, and (2) the tank had to have features that resisted impacts from vehicle collisions and bullets. The appendix item gave local jurisdictions the option to adopt or disregard the new code language for aboveground fueling tanks [13].

In addition to activity among code-making bodies, there were new environmental-protection interpretations that would allow ASTs without normal dikes typically associated with aboveground tanks. The EPA also had a rule that covered ASTs, which was developed in 1972 as part of the Clean Water Act. According to the Act, all aboveground storage tanks that contained fuels and chemicals had to provide diking as a means of containment in case of a release, and to prevent the flow of hazardous liquids into navigable waterways.

In 1992, EPA issued an interpretation that allowed secondary containment double-wall tanks to be installed without diking under certain conditions, which included a tank capacity below 12,000 gal and the use of overfill prevention equipment [14]. In 1993, the NFPA 30 code issued a similar change. The fire codes required diking or remote impounding to assure a release would not fuel a fire at another location. The revision allowed secondary containment tanks as an exception to the spill control requirements. Again, a number of requirements to satisfy this exception were established, such as a maximum 12,000-gal capacity, openings at the top of the tank, emergency venting of the interstice, overfill shutoff devices, overfill prevention alarms, and gages for transport driver use. The main goal was to prevent common causes of AST system releases. Some of the provisions included in the NFPA 30 code paralleled the UFC requirements [15]. For service station tanks, NFPA 30A (Automotive and Marine Service Station Code) took this further by mandating security in the form of fencing and gates to prevent vandalism of thc tank. Also required were antisiphon devices [16].

All of these changes paved the way for more secondary containment tanks. By 1993, STI members were building more aboveground tanks than underground tanks, which reflected a dramatic change in the market. By the mid-1990s, STI members were building two or three ASTs for every UST. A survey of tank owners published by STI in 1993 verified this trend. The survey results indicated that most tank owners were concerned with overfills and vandalism as the two greatest threats for AST releases, and that fire codes had the most authority over such tanks [17].

The EPA was also active during this time span on aboveground tanks. A major spill in Pittsburgh from the failure of a large field-erected tank, which caused pollution of the Ohio River for hundreds of miles, brought national attention to ASTs in 1988. Further, a slow but constant release in Fairfax, VA, caused flammable fumes to migrate into an affluent neighborhood and grab the attention of the Congress. As a result, several U.S. senators and representatives began to pursue legislation that would increase regulation of ASTs. However, none of the law-making initiatives garnered much popular support.

In 1991, EPA proposed changes to the Clean Water Act, under the Spill Prevention Control and Countermeasures (SPCC) program. The regulatory changes strengthened the language in federal policy, changing "shoulds" to "shalls," while adding a proposal for integrity testing of tank and pipe systems. It also proposed that ASTs employ diking that would be impermeable for 72 hours. This proposal, if it had been adopted, essentially would have eliminated most earthen dikes from regulatory compliance. As of this publication, several other revisions and additions had been proposed to the SPCC regulations, but the final rules had not been released.

The shop-fabricated AST market saw a major change in customer demand. Specifiers were asking for steel dikes or tubs within which a tank would be installed. Steel, an impermeable material, certainly met the requirements of the EPA proposal. But demand for integral double-wall ASTs also was growing.

So between the fire codes and the environmental regulation proposals, the demand for secondary containment tanks increased exponentially. Prior to 1990, a shop-built secondary containment AST was rare. By the mid-1990s, it became the norm—accounting for at least a third of STI members' AST sales.

National standard development work paralleled these significant changes. Underwriters Laboratories in 1993 nearly tripled the size of its UL 142 tank standard document by adding provisions for steel dikes, double wall tanks, and rectangular tanks. Rectangular tanks became very popular for ASTs smaller than 1,000 gal. Because of their flat-top design, the rectangular models made it easy to access the top of small tanks where fittings and fueling equipment could be located. STI also developed several standards that addressed changes in marketplace demand. In 1991, STI issued a standard for a single-wall AST contained in a steel dike. One year later, STI introduced its dual wall AST standard, F921. Both construction standards met UL 142 requirements.

But that was not the end of new standard development work.

1. NFPA 30A in 1993 allowed UL 142 tanks to be installed, as well as fire-resistant tanks. The new criteria in NFPA 30A for fire-resistant tanks said the tank construction must prevent release of liquids, failure of the supporting structure, and impairment of venting for a period of not less than two hours when tested using a fire exposure that simulates a high intensity pool fire, such as that described in UL 2085 (a draft of which was available for code officials' review during 1993) [16].
2. The UL 2085 standard was published in December 1994. It provided two listings, a fire-resistant tank and a protected tank. The fire-resistant listing required exposure to a two-hour, 2000°F test. However, during that test, the internal tank temperature change could not exceed 1000°F. In addition, UL 2085 provided a listing for a protected tank, which prevents an internal tank temperature increase of more than 260°F when exposed to a 2000°F pool fire. This listing essentially duplicated the requirements within the Uniform Fire Code Appendix II-F. UL called the standard Insulated Tank for Flammable and Combustible Liquids [18].

Code development work continued to accelerate at rapid pace. Revisions were made to the UFC Appendix II-F each year throughout the 1990s and to the

NFPA 30A Code in its 1996 edition. In addition, the Building and Code Administrators International (BOCA) and Southern Building Congress Code International (SBCCI) had fire codes that also brought forth changes to allow aboveground storage tank systems for motor vehicle fueling facilities. However, in the case of BOCA and SBCCI, an AST system could not be installed at retail service station facilities [19,20].

By 1997, STI had developed a statistically valid database of insulated protected tanks and double wall F921 tanks. The insulated tank was called Fireguard—the specification for which was published in 1994. STI's data showed that less than 3 percent of these ASTs were installed at public retail service station facilities. Private fleet facilities were the primary and dominant users of aboveground tanks for fueling vehicles. Nearly two-thirds of stored-product applications were for less flammable Class II or III liquids, such as diesel, kerosene, or lube oils.

Because of the ongoing code changes, Underwriters Laboratories made further modifications to its UL 2085 standard. In December 1997, UL removed the fire-resistant tank listing provisions from UL 2085 and moved it to a new standard, UL Subject 2080. Appropriate titles to reflect the type of tank were assigned to each standard [21]. UL also issued a standard for tanks installed in vaults. The UL 2245 standard covers below grade vault construction, design, and testing.

Finally, authorities having jurisdiction continued to have concerns about important pipe fittings that were missing from ASTs upon installation, such as emergency vents. In 1997, UL introduced its AST system listing under a new standard, UL 2244. In this case, all important fittings and pipe attachments would be either shipped on or with the aboveground tank.

VIII. SAFER STORAGE SYSTEMS

From 1988 to 1998, the underground tank population in the United States fell by more than 55 percent. The number of retail service stations also dropped dramatically from over 207,000 in 1992 to less than 183,000 in 1998 [22]. Consolidation among oil companies was continuing. Tanks were getting larger as the throughput of fuel at each service station increased. More compartment tanks were being installed. The market had changed so dramatically during the decade that even prefabricated tank system designs (tanks prefabricated with piping) had begun to appear in the underground tank market.

Petroleum and chemical storage was far advanced from the tubs, wash basins, and whiskey barrels rushed into use in Titusville when the first drill found rock oil [1]. Because of public safety and environmental concerns, the storage of flammable and combustible liquids is handled today with unprecedented attention to ensuring that product remains in reliable containment.

REFERENCES

1. D Yergin. The Prize. New York: Simon & Schuster, 1991, p 27–28.
2. National Board of Fire Underwriters. Rules and Requirements of the National Board of Fire Underwriters for the Construction and Installation of Systems for Storing 250 Gallons or Less of Fluids Which at Ordinary Temperatures Give Off Inflammable Vapors, as Recommended by its Committee of Consulting Engineers. NBFU 30-1904 standard, 1904.
3. Steel Tank Institute, sti-P_3® Specification and Manual for External Corrosion Protection of Underground Steel Storage Tanks. Lake Zurich, IL, STI-P3-97 specification, 1997.
4. National Fire Protection Association. Flammable and Combustible Liquids Code, Quincy, MA, NFPA 30 Standard, 1977, 1981.
5. Steel Tank Institute. Standard for Dual Wall Underground Steel Storage Tanks. Lake Zurich, IL, F841-98 Standard, 1998.
6. National Association of Corrosion Engineers. Control of External Corrosion on Metallic Buried, Partially Buried, or Submerged Liquid Storage Systems. Houston, TX, RP 0285-85 Recommended Practice, 1985.
7. Versar, Inc. Summary of State Reports on Releases from Underground Storage Tanks. Washington, DC, Report to U.S. Environmental Protection Agency, 1986.
8. JH Pim, JM Searing. Final Report: Tank Corrosion Study. Suffolk County, NY. Department of Health Services. Farmingville, NY, Report to U.S. Environmental Protection Agency, 1988.
9. F Mosely. Gasoline and Groundwater Contamination: A Background Paper Prepared for Northeast-Midwest Senate Coalition, 1985.
10. Ohio Environmental Council. 20,000 Leaks Under Ohio: Environmental and Health Threats from Leaking Underground Storage Tanks in Ohio. Columbus, OH, 1988.
11. Federal Register. Underground Storage Tanks: Technical Requirements and State Program Approval (Final Rules) Part II, 40 CFR Parts 280 and 281. 1988.
12. Steel Tank Institute. Standard for STI-86 Containment System. Lake Zurich, IL, F861-86 Standard, 1986.
13. International Fire Code Institute. Uniform Fire Code, Appendix II-F. Whittier, CA, 1993.
14. DR Clay. Use of Alternative Secondary Containment Measures at Facilities Regulated under the Oil Pollution Prevention Regulation (40 CFR Part 112). Washington, DC: U.S. Environmental Protection Agency memorandum, 1992.
15. National Fire Protection Association. Flammable and Combustible Liquids Code, Quincy, MA, NFPA 30, 1993
16. National Fire Protection Association. Automotive and Marine Service Station Code. Quincy, MA, NFPA 30A Code, 1993.
17. Product Evaluation, Inc. Aboveground Storage Tank Boom—Will the Trend Continue? Lake Zurich, IL, Survey for Steel Tank Institute, 1993.
18. Underwriters Laboratories. Protected Aboveground Tanks for Flammable and Combustible Liquids. Northbrook, IL, UL 2085 Standard, 1994.

19. Building Officials & Code Administrators International, Inc. The BOCA National Fire Prevention Code. Country Club Hills, IL, 1996.
20. Southern Building Code Congress International, Inc. Standard Fire Prevention Code, 1996.
21. Underwriters Laboratories. Outline of Investigation for Fire Resistant Aboveground Tanks for Flammable and Combustible Liquids. Northbrook, IL, Subject 2080 Standard, 1997.
22. Association of State and Territorial Solid Waste Management Officials. Report Card on the Federal UST/LUST Program. Washington, DC, 1998.

2
Historical Perspective on Standards and Codes

Marshall A. Klein
Marshall A. Klein & Associates, Inc., Eldersburg, Maryland

Industries that require the storage of flammable and combustible liquids face a complex array of codes and standards with which they must comply. Besides adhering to environmental regulations, underground and aboveground tank systems must be sited and operated in accordance with local building and fire codes.

Local codes, mostly, reflect the work done by national or regional code-making bodies. A city council or county board may adopt for their jurisdiction a nationally developed code—and add some local distinctions as well. This chapter will provide some historical background on the regulatory function of codes and standards, and cite some of the codes that most affect storage tank systems.

I. WHAT IS A STANDARD?

A standard is a series of requirements that tell you *how* to do something. A standard tends not to have any enforcement requirements. A standard becomes an enforceable document when it is adopted by reference in a code [1].

II. WHAT IS A CODE?

A code is a set of regulations that tells you *when* to do something. A code will have requirements specifying the administration and enforcement of the document.

A. What is a Building Code?

A building code is a set of regulations legally adopted by a community to ensure public safety, health and welfare insofar as they are affected by building construction.

B. What is a Fire Code?

A fire code is a set of regulations legally adopted by a community that define minimum requirements and controls to safeguard life, property, or public welfare from the hazards of fire and explosion. A fire code can address a wide range of issues related to the storage, handling or use of substances, materials or devices. It also can regulate conditions hazardous to life, property, or public welfare in the occupancy of structures or premises [1].

III. ORIGIN OF BUILDING AND FIRE CODES

The writing, adoption, and enforcement of codes have a centuries-old tradition. For at least 5000 years, man has exercised some limited controls over the construction and utilization of buildings and structures throughout the civilized world.

A. Historical Perspective of Codes

1. 2000 BC—Hammurabi

Hammurabi, founder of the Babylonian Empire and known as an effective leader, wrote the earliest known code of law. Portions of the Hammurabi Code of Law dealt with building construction. Here are some translated examples of Hammurabi's Code of Law on Building Construction:

> No. 229: If a builder has built a house for a man and his work is not strong, and if the house he has built falls in and kills the householder, that builder shall be slain.
>
> No. 230: If the child of the householder is killed, the child of that builder shall be slain.
>
> No. 231: If the slave of the householder is killed, he shall give slave for slave to the householder.
>
> No. 232: If goods have been destroyed, he shall replace all that has been destroyed; and because the house was not made strong, and it has fallen in, he shall restore the fallen house out of his own material.
>
> No. 233: If a builder has built a house for a man, and his work is not done properly and a wall shifts, then that builder shall make that wall good with his own silver.

In Polynesia, an even sterner law than the Hammurabi Code of Law required builders to entomb a live slave under each corner post, which was meant to guarantee that the structure would be supported properly, in perpetuity.

2. 64 AD—Burning of Rome

In 64 AD, Emperor Nero had a master plan for construction of a new Rome prepared sometime prior to the fire that destroyed much of the city. Prior to Nero's ascent to power, Rome lavished its wealth and resources on the construction of public buildings, but ignored almost all other construction regulation. Many of these buildings collapsed even before they were completed, killing and maiming many of the workmen. Consequently, the charges that Nero deliberately ordered Rome's destruction seem to be well founded. The rebuilding of Rome was accomplished in accordance with sound principles of construction, sanitation, and utility. Until the downfall of the Roman Empire, construction, public and private, was closely monitored and controlled. In hindsight, it could be stated that Rome's burning might have been the world's first urban renewal program.

3. 798, 982, 1212, 1666—Fires in London

London had major conflagrations in 798, 982, and 1212. The fire in 1666 is singled out because it resulted in the British government's adoption of building controls.

London was destroyed in the great fire of 1666. Destruction may have been more of a blessing than a calamity because London was a crowded, filthy city of low, timber-framed warehouses, churches, and houses. Most of its thoroughfares had open drains that carried raw sewage. Housewives threw garbage into narrow cobblestone streets. London had been ravaged by the plague for nearly a year prior to the fire. Its people were dying at the rate of 1000 per week. The fire started in a rundown neighborhood near the Tower of London. The fire raged for five days and nights. Its toll was 15,000 buildings, including 84 churches. Miraculously, only six lives were lost in accidents directly attributable to the fire.

In its attempt to prevent another devastating fire, Parliament decided to write requirements to control building construction. However, it took Parliament two years to enact such controls for buildings. The regulations were called the "London Building Act," and it applied only to property within the boundaries of London. Unfortunately, during those two years, London was reconstructed almost in the same style that existed prior to the fire.

4. Major Fires in the United States

From 1800–1900, 11 major American cities were devastated by fire. The following two examples of conflagrations in cities in the United States show the importance of building regulations:

a. 1871—Chicago Fire. The Chicago fire in 1871 was devastating, and the second most costly blaze in American history. Chicago at the time consisted of about 60,000 buildings, more than half of which were of wooden construction. Lloyds of London had warned its underwriters in Chicago of the city's conflagration potential, but the insurance companies continued to issue fire insurance coverage.

The initial fire, which by legend was blamed on Mrs. O'Leary's cow kicking a lantern, started on October 7, 1871. Local officials thought after a few hours that the fire posed no more danger. However, on the night of October 8, a new fire broke out and, fanned by winds whipping off Lake Michigan, was soon raging out of control. The fire grew to such large proportions that drastic measures were employed by the army, including the use of explosives to create firebreaks.

Before the fire was extinguished two days later, 17,000 buildings had been destroyed and 250 lives had been lost. Almost 100,000 persons were homeless. Fire damage was $168 million in 1871 dollars. In 1994 dollars, the fire damage would be more than $2 billion. Sixty insurance companies went into bankruptcy because of fire-related claims. The insurers that survived the financial disaster threatened to leave the city unless adequate laws regulating buildings were enacted. In 1875, a building code and a fire prevention ordinance became effective in Chicago.

b. 1906—San Francisco Earthquake and Fire. The San Francisco fire that started on April 18, 1906, was the largest loss fire in U.S. history. Fire damage was $350 million in 1906 dollars. In 1994 dollars, the fire damage would be $5.7 billion.

B. What is the Purpose of Controls, and Why do We Need Building and Fire Codes?

As can be seen, building regulations, as we know them today, are the result of an evolutionary process that has its roots deeply embedded in disaster and tragedy. The absence of controls, and the absence of enforcement, must share the responsibility for the needless loss of lives and property. It would be proper and safe to say that in the past 5000 years, millions of lives have been sacrificed for lack of such laws [1].

IV. MODEL BUILDING AND FIRE CODE GROUPS IN THE UNITED STATES

The following organizations publish model building and fire codes that jurisdictions in this country have adopted by reference.

Building Officials and Code Administrators International, Inc. (BOCA)
Southern Building Code Congress International, Inc. (SBCCI)
Southeast Association of Fire Chiefs and Southwest Association of Fire Chiefs
International Conference of Building Officials (ICBO)
International Fire Code Institute (IFCI)
National Fire Protection Association (NFPA)

A. Brief History of the Model Building and Fire Code Groups in the United States

1. NFPA

Founded in 1896, NFPA published its first code, an edition of the sprinkler code, in the same year as the organization's inception. Other codes of interest to the storage tank industry are NFPA 30 (Flammable and Combustible Liquids Code), and NFPA 30A (Automotive and Marine Service Station Code). From 1913 to 1957, the NFPA 30 Code was known as the Suggested Ordinance for the Storage, Handling, & Use of Flammable Liquids. Prior to 1984, NFPA 30A originated as Chapter 7 of NFPA 30. NFPA's national fire codes and standards are developed and revised through the deliberations of 235 technical committees. These committees are composed of a balanced membership of technically competent individuals from industry, government and independent experts. One-third of these codes and standards are revised every year. All NFPA Codes and Standards are usually on a three-year revision cycle unless the committee for a code or standard requests an extension of a year or two.

All dues-paying members of NFPA can vote on the adoption of codes and standards at the association's fall and spring meetings.

2. BOCA

BOCA was founded in 1915. BOCA published its first building code in 1950; it was called the BOCA Basic Building Code. Today, the building code is called the BOCA National Building Code. BOCA published its first fire-prevention code in 1966, the BOCA Basic Fire Prevention Code. The fire prevention document today is known as the BOCA National Fire Prevention Code.

The BOCA code change cycle covers three years. Here's a typical sequence of events in the cycle: The 1996 editions of the BOCA national codes were issued during that year. (The year the full edition of the Code is published serves as a moratorium period for any code changes.) But in 1997 there were meetings to discuss code changes, which led to the 1998 Supplement to the BOCA National Building Code/1998 and BOCA National Fire Prevention Code/1996. In 1998, more code changes were proposed and voted upon, which resulted in publication of the revised 1999 BOCA national codes.

Only code-official members can vote on changes to the BOCA national codes.

3. SBCCI

Founded in 1940, SBCCI published in 1945 the first edition of the Standard Building Code. SBCCI published in 1974 the Standard Fire Prevention Code. Since 1991, the Standard Fire Code has been under the control of the Southeast Association of Fire Chiefs and Southwest Association of Fire Chiefs. The SBCCI code cycle covers three years. SBCCI issues code supplements for the first two years. A supplement is issued for each year; the second year's supplement is cumulative—including changes from the previous supplement. Then, in the third year, SBCCI issues completely revised editions of its codes.

Only code-official members can vote on changes to the standard codes.

4. ICBO

Founded in 1922, ICBO published in 1927 the Uniform Building Code. ICBO published its first fire code in 1971, the Uniform Fire Code. From 1971–1991, the Uniform Fire Code was under the control of the Western Fire Chiefs Association. In October 1991, the International Fire Code Institute (IFCI) was formed to manage the Uniform Fire Code.

The last edition of the Uniform Building Code was published in 1997. The new International Building Code (IBC), which is scheduled for publication in 2000, will not only replace the Uniform Building Code, but also the BOCA National Building Code and the Standard Building Code. The IFCI will continue to publish and maintain the Uniform Fire Code with supplements in 1997 and 1998, and a completely revised edition in 1999.

Only code-official members can vote on changes to the Uniform Building Code and the Uniform Fire Code.

a. Formation of the ICC. On December 9, 1994, the three model building code groups (BOCA, SBCCI, and ICBO) formed an umbrella organization to support common code development. They called it the International Code Council (ICC). Its mission is to promulgate a single model code system to meet global economic trends.

B. How Does a Code or Standard Become Enforceable?

A code or standard becomes enforceable when it is adopted by reference through local, state, or federal government legislative process, such as an ordinance, statute, or bill. The law must adopt a specific edition (year of publication) of a code or standard, and may include amendments to specific portions of the code or standard being adopted.

REFERENCE

1. RL Sanderson. Codes and Code Administration: An Introduction to Building Regulations in the United States. Chicago: Building Officials Conference of America, 1969.

3
History of the Uniform Fire Code

Lavern Cary
Cary & Associates, Sherwood, Oregon

I. THE LONG EVOLUTION

As noted in Chapter 2, fire codes in use in the United States have an interesting history. One of the most prominent codes in modern use is the Uniform Fire Code, which regulates underground storage tanks, aboveground storage tanks, and a vast array of products and services that could affect public safety. The Uniform Fire Code, which is cited within the ordinances of communities in 38 states, is like many other codes. The current version is the result of an evolutionary process.

Today's codes are essentially the refinement of work started during the mid-1800s by an organization—the National Board of Fire Underwriters (NBFU)—to serve the interests of fire insurance companies. That organization developed and first published in the early 1900s a model code called the National Fire Code.

In the early 1900s only a few large cities had comprehensive codes. Many of those codes were based upon the model National Fire Code. A few cities actually took it upon themselves to develop and adopt a code unique to their communities.

Some state governments created code enforcement agencies, usually as part of the state insurance commissioner's office, or an independent state fire marshal's office. There was a wide variance as to what each state adopted for a code. Many adopted, or simply used, some or all of the National Fire Protection Association (NFPA) pamphlets, which covered a broad spectrum of fire safety subjects.

II. INFLUENCE OF WORLD WAR II

The greatest early influence that shaped changes in fire codes was World War II. As industry rapidly ramped up for war production—and relatively large segments of the population migrated to provide labor for those industries—significant societal changes were unleashed. Public safety officials became concerned with how to manage population growth, urban density, new materials and processes of a hazardous nature, and transportation issues. That influence carried over to the postwar era, especially in states such as California, which experienced unprecedented growth that compounded the region's problems related to fire safety.

The postwar era led to a realization by officials in smaller cities of the value and necessity for adopting and enforcing a fire code. That necessity was emphasized to local decision-makers by the activities of the Insurance Rating Bureau, another organization whose function was primarily to look out for the interests of insurance companies. Communities were encouraged to adopt the National Fire Code. Around this time the code publishing organization's name also evolved—from the National Board of Fire Underwriters to the American Insurance Association.

As more communities adopted codes, enforcement personnel became more experienced, organized, and sophisticated. Recognition was growing that the National Fire Code was focused primarily on property protection. It was generally lacking in regulations aimed at the protection of life. Further, there was a generalist approach to the document. Communities that were home to industries with special hazards were forced to be creative in amending the code to provide for their unique needs.

NFPA offered a solution for some of those needs. They continued to expand the selection of fire safety pamphlets and standards, giving communities choices that were not earlier available. One of the NFPA documents that became widely used was the Safety to Life Code. It was focused on assuring that people were protected, not just property. Its earliest editions were generally referred to as the NFPA Exits code. The thrust was to assure that buildings were designed safely enough that occupants could exit in the event of a fire.

Meanwhile, building codes were evolving and becoming more widely adopted. Nevertheless, it is interesting to note that as recently as the 1950s a large number of cities and counties had no comprehensive building codes. Building code adoption at the state level was not common until even later. Even today, there are areas of the country that do not adopt or enforce building codes, which typically are written and dominated by code enforcement officials with input and support from architects and representatives of the building materials industry.

III. RESOLVING INEVITABLE CONFLICTS

With the confluence of the heavily amended National Fire Code, the NFPA Safety to Life code, locally created fire codes, and model building codes, a new set of problems surfaced. There were conflicts between the requirements of the building code and fire code adopted by individual communities. Those conflicts created problems of enforcement between building and fire code personnel. Architects, builders, building owners and operators, and many other stakeholders were constantly caught in the middle of turf battles.

A case in point, in the extreme! A large apartment building was constructed—allegedly in compliance with the building code—and given a certificate of occupancy by the building official. Once it was opened, the local fire official inspected it for compliance with a conflicting fire code, found it in noncompliance, and ordered the building demolished. When the case went to court, a judge upheld the fire official and a brand-new building was knocked down. While that was an unusual case, the problems of conflict persisted to an intolerable degree.

Determined to solve the problem, the California Fire Chiefs Association (CFCA) and the International Conference of Building Officials (ICBO), based in California, started a dialogue. Out of that came the recognition that a solution should be more broadly based than the concerns raised in just one state.

Around 1969 conversation started between the ICBO, writers of the Uniform Building Code, and the Western Fire Chiefs Association (WFCA) regarding ways in which code conflicts might be addressed and mitigated. Both of those organizations were based in the western states. Party to those conversations were CFCA representatives, who had already developed an expanded fire code based upon the National Fire Code. Out of those conversations came agreement to develop—jointly and contractually—a new Uniform Fire Code based upon the California Fire Chiefs code.

IV. THE NEW CODE

In 1970 the two organizations formed a work committee of eight officials—four from fire code enforcement, and four from building code enforcement. Their charge was to create a draft fire code, carefully reviewed and amended to eliminate, as nearly as possible, conflicts between the Uniform Building Code and the proposed fire code. Staffing for the effort was through the International Conference of Building Officials. By contract, the document copyright was to be owned on a 50/50 basis. The future development of the code would be by the WFCA. The document staffing and publication responsibilities would remain with the ICBO.

The development of the first 1971 Uniform Fire Code draft consumed the better part of the year. The WFCA at their annual meeting considered and voted to approve the document. The work committee was dissolved. The code was published. A new WFCA code development committee was appointed to assume the task of conducting code change hearings on an annual basis. The next code was published in 1973. Since then a new edition has been published every three years with supplements printed in the off-publication years.

V. A NEW ORGANIZATION

Over time, recognition grew that activities associated with the Uniform Fire Code should be expanded. For example, the consensus was that training and inspector certification programs would be beneficial. There was also an increased desire to broaden the use and influence of the Uniform Fire Code from its original western states base to a national constituency. In 1991 the WFCA and the ICBO jointly created a new umbrella organization named the International Fire Code Institute. Its board was made up of persons from both organizations, as well as individuals from other code groups throughout the United States. Its purpose was to assume all of the responsibilities for administration, code development, training services, and other related activities previously carried out jointly by WFCA and ICBO. The Uniform Fire Code copyright remained with the originating organizations. Staffing remained with the ICBO.

The first Uniform Fire Code edition published by IFCI was 1991. The IFCI continues to operate today. The most recent Uniform Fire Code edition is 1997 with supplements issued after the annual meeting in the summer of 1998.

As the move toward an international fire code has gained momentum, the Uniform Fire Code's future has been a source of controversy. The IFCI intended to cease oversight and publication of the Uniform Fire Code by the year 2000. However, at the time of this book's publication, the Western Fire Chiefs Association had announced an intent to continue development of the Uniform Fire Code through annual code cycles into the next century.

4
Quality Control on USTs and ASTs

Larry O'Shea
Steel Tank Institute, Lake Zurich, Illinois

I. OBTAINING A QUALITY PRODUCT

Steel tanks fabricated in the 1950s and 1960s are not much different from the tanks of today. Yes, it's true that advancements in corrosion protection and secondary containment have remarkably changed products built to Underwriters Laboratories UL 58 standard for underground storage tanks (USTs). And it's true that extraordinary innovations in aboveground storage tank (AST) design have forever altered the concept of what a UL 142 tank would look like.

But for 98 percent of USTs and ASTs manufactured, when you strip away the latest product features, you still are working with a steel cylinder equipped with a few fittings that will be used for the storage and handling of petroleum or chemical products. In the remaining cases, you are likely to be working with a flat-top rectangular tank—most often storing less than 2000 gal.

Many companies are qualified to fabricate steel ASTs and USTs, but it's the quality of their effort that ultimately will make the biggest difference in how well the tank serves the end user. From a quality control perspective, tank buyers should be aware of essential fabrication requirements to ensure receipt of a quality tank that will provide many years of trouble-free service. These essentials would include quality control (QC), design, materials, tank suppliers, fabrication and welding, surface preparation, coatings, and inspection. Equally important are shipping, handling, and installation. Because of some key differences in the installation of ASTs and USTs, see Chapters 13 and 23 for more information. Each of these areas should be considered prior to ordering a tank.

II. QUALITY CONTROL

It is imperative in today's litigious world to approach the specification and purchase of storage systems with quality in mind. Purchase orders should concisely identify materials, specifications, inspection/testing requirements and documentation. Buyers should evaluate a supplier's capabilities and examine the workmanship of existing orders. Does the supplier have a QC program with qualified inspectors and well-documented procedures? Even in the best QC programs mistakes will happen; however, a good manufacturer will identify the problem, make the necessary correction and try to prevent the situation from reoccurring on the next tank.

III. MATERIALS

Since tanks have relatively few components, materials are fairly easy to monitor—compared to a product such as an automobile. Most tanks are made from commercial-quality hot-rolled steel. Many tank fabricators will purchase steel meeting ASTM requirements, for example A36 or A569. More important to the steel, whether it be certified ASTM, or commercial quality, is the carbon or carbon equivalency content. The carbon content of the steel should not exceed 0.3 percent. In addition, the carbon equivalency should not exceed 0.53 percent. To calculate the carbon equivalency (CE) of steel, use the following formula:

$$CE = C + (Mn + Si)/6 + (Cr + Mo + V)/5 + (Ni + Cu)/15$$

Increased amounts of carbon reduce the weldability of the steel and increase the hardness and tensile strength, which can make the steel difficult to work with.

Other materials that are very important and will be discussed in greater depth are the welding electrode (or filler material) and the corrosion protection coatings. The welding electrode/filler material is critical, since many tank failures develop at the weld seams.

IV. FABRICATION AND WELDING

Tank fabrication is very labor intensive, which means the skill of a craftsman is very evident to a trained eye. The steel is usually purchased in coil form before it is flattened and cut to length. Many fabricators will purchase their steel from a supply house that will flatten and cut the steel to length prior to shipment.

At the tank plant, the steel is rolled into cylinders and tack welded before it's moved to a welding station. Typically, the long seam of the cylinder is either butt welded or lap welded. In many instances only the inside surface is welded prior to further assembly. Many fabricators will use what is called an offset (or joggle) joint to connect the cylinders or cans together. The joggle joint has a couple of advantages. Because there is an offset, the assembly and fit-up of one can—or course of steel—to another is relatively easy. Also, the resultant weld joint is relatively consistent, making it desirable for automatic welding processes.

Most fabricators will use a flat, flanged head on their tanks. Flanged heads are often purchased from outside suppliers, although just about every fabricator has a flanger on site for making heads. All flanged heads should have an inside knuckle radius of two times the thickness of the head. Because the knuckle is cold-worked—meaning it hasn't been heat-treated to relieve stresses—it is a natural stress point on the tank. Steel Tank Institute's quality inspectors, for example, closely monitor the knuckle radius for signs of proper fabrication.

Most fabricators will use a single piece of steel for making a head, whenever possible. However, because of the diameter required on larger tanks (12,000 gal capacity or greater), stock materials matching the diameter are often unavailable. When this happens, a fabricator will have to piece together two or sometimes three sheets of steel to meet the required dimensions. The fabricator will then weld the sheets together and cut them to match the radius required for the tank. Prior to flanging the head, the fabricator will grind the weld flush with the base metal in the areas that will come in contact with the flanger die. The stress created at the weld during the flanging operation can feasibly produce cracks.

Typically, most tank interior welding is completed prior to fitting the last head on the tank. Many tanks now have manways for interior access, but most fabricators will try to minimize their employees' exposure to working in confined spaces. After the final head is fit and tack welded into place the tank's exterior seams are welded. Then the openings are cut for placement of the fittings, or other attachments. Once the tank is complete, welds will be cleaned and prepared for leak testing.

Of course, not all facilities will follow this procedure. As a matter of fact, it is very rare to find any two fabricators building tanks in an identical manner because every production shop has found special success formulas that make them unique.

A. How Tanks Are Welded

Four basic welding processes are used in fabricating steel tanks and pressure vessels. Each of the four processes (whether manual or automated) will repeatedly produce quality weldments. But they each have strengths and weaknesses. Prospective tank buyers should be aware of the different processes. In general, the best fabricators focus intensely on assuring outstanding fit-up. Many combine that dedication with the use of automatic weld processes—which typically leads to high-quality welds.

SMAW (Shielded Metal Arc Welding)

Commonly referred to as stick welding, SMAW is the most commonly used welding process. Equipment and materials are inexpensive, portable, and relatively simple to operate, even outdoors.

However, the SMAW process is slow. Because it is a manual process, the operator must stop and replace the consumed electrode approximately every 18 in. In addition, the operator must remove slag from the weld surface. SMAW welds are readily identifiable by start/stop markings that lead to a greater number of surface irregularities than would be found with other processes.

The type of electrode used in the SMAW process has a significant impact on the mechanical strength created by the weld. For example, a jet rod (known for its speed) will not penetrate as easily into the base material; thus, the strength of the joint is not as great as other types of electrodes, such as pipe rod.

GMAW (Gas Metal Arc Welding)

Also known as metal inert gas (MIG) welding, the GMAW process uses a solid bare wire to create the arc between the work piece and the filler material. The wire is fed (usually automatically or semiautomatically) through a welding gun from a spool or reel. A shielding gas protects the weld from environmental contamination (principally oxygen). The gas flows through the welding gun/nozzle to the arc.

The GMAW process has a high deposition rate (i.e., the rate at which the electrode melts and gets deposited on the work piece) and can save time and labor costs. However, because the process uses shielding gas, it is very susceptible to drafts or wind that might blow the shielding gas away from the arc. Thus, it is not well suited for welding outside.

While GMAW welds generally have a good uniform profile, MIG welding is less forgiving than some of the other processes. A greater risk of lack of fusion exists because the arc doesn't provide as much heat. Many fabricators will limit its use to thinner gauge steel. Undesirable weld spatter (metal particles expelled during welding that do not form a part of the weld) is generally greater with this process, but can be significantly reduced with the proper mixture of gas and the use of anti-spatter material on the tank surface.

FCAW (Flux Core Arc Welding)

The FCAW process is very similar to MIG welding, although it does not always need shielding gas. Also, instead of a bare wire, a tubular wire containing granular flux is used. When a self-shielding flux is used, no shielding gas is required.

Some wires use a combination of both flux and shielding gas for improved weld metal properties.

The FCAW process is very popular because of the deep-penetrating arc that reduces fusion type discontinuities, and because it has a quick deposition rate. The self-shielding electrode is popular in field applications as well. Like the SMAW process, solidified slag must be removed from the surface of the weld.

The FCAW process is known for generating smoke, which often hangs in the air of a manufacturing plant. Although similar to MIG welding, the FCAW process often presents a different kind of problem: slag. Because the tubular wire has flux inside, there is greater opportunity for slag entrapment in the weld. Undercutting (melting away of the base material along the edge of the weld forming a groove) is very common with this process because the arc creates significant heat.

SAW (Submerged Arc Welding)

SAW is probably the most efficient welding process because it's automatic and uses a solid wire electrode submerged in a granular flux blanket. As with the FCAW process, SAW has a deep penetrating arc, and solidified slag must be removed from the surface of the weld. The deposition rate of the SAW process is excellent, especially on thicker materials requiring multiple passes.

The SAW process produces a very smooth and uniform weld, but it can also pose problems. With the automatic process the tank must be rotated at a constant speed, and the weld joint fit must be uniform. With other processes (manual and semiautomatic), the welder can make adjustments to conform to the joint geometry. With SAW, the welder cannot see the arc and is limited to only minor adjustments while the unit is in operation. Again, slag entrapment and fusion discontinuities are fairly common with this process. Also, because of the amount of heat at the arc, some caution is important—many SAW operators have burned right through the base metal.

When evaluating a tank supplier's welding program it's important to verify that the supplier has qualified procedures and operators. Good tank suppliers will have welding procedure specifications (WPS) for each process. The WPS will identify the volts and amperages, electrode/filler material, shielding gas, etc., to produce a quality weld. A fabricator will have visual, mechanical, and sometimes nondestructive tests performed on weld samples to qualify a WPS or a welding operator.

V. LEAK TESTING

Once a tank is completely welded, the tank will be visually inspected and checked for leaks. Most cylindrical atmospheric tanks are tested at 5 psi. Tank openings are sealed, and compressed air is applied to the tank until the necessary test pressure is achieved. It is important to consider the volume of air applied to the tank when pressurizing. While the test pressure is not that great, the volume of air is significant and, in the event of an accident, can cause severe injury to nearby personnel. Fabricators are required to have a safety relief device on the tank to prevent overpressurization.

Once the test pressure is achieved, a soap solution is applied over the welds to identify leaks. Bubbles are created when air leaks through the weld and are readily identified by the tester. Leaks are then repaired before the welds are retested to ensure the tank's integrity.

Testing of tanks should be performed in a well-lit area. All gages, relief valves, and testing equipment should be in good working condition. Soap testing solution should be protected from contaminants. The tank welds should be free of slag, spatter, or foreign materials that prohibit testing.

Most double-wall tanks will, in addition to the bubble test, have a vacuum applied to the interstitial space. Comparing a vacuum to the bubble test is like comparing apples to oranges. The vacuum is much more sensitive to leaks than the bubble test, and is most effective at proving simultaneously that both the primary and secondary tanks are tight.

Some distinctions exist for testing on various UST and AST technologies:

All single-wall and double-wall steel tanks should be tested with an air test that includes soaping of the weld seams to look for pinhole leaks—some fabricators recommend an additional vacuum test on Type I (blanket wrap design) double wall tanks.

Jacketed tanks typically are shipped with vacuum, but if a leak develops in the jacket the external shell could be tested with an ultrasonic device or a helium leak detector.

Steel dikes on AST systems can be examined using a dye penetrant, or penetrating oil, on weld seams.

VI. SURFACE PREPARATION

Tanks that have coatings or laminates applied directly to the exterior steel shell should be prepared by abrasive blasting, which will remove mill scale and roughen the surface to improve adhesion of the paint or coating. Typically the resultant profile left on the steel is between 1.5 and 3.5 mils deep.

The blast medium used is often coal slag, silica sand, or recycled grit. The profile depth depends upon the size, type, and hardness of the blast media; particle velocity; angle of impact; surface hardness; and maintenance of the working mixture.

The coal slag and grit tend to give a deeper profile versus the sand. However, the sand tends to remove mill scale easier from the surface. Under most circumstances shot or wet blast should not be used for surface preparation on steel tanks.

Prior to coating, the tank should be free of abrasives, oil, grease, or other contaminants. Most fabricators will apply the coating to the tank as soon as possible after blast. The longer a blasted, uncoated tank is exposed to the environment, the greater the chance that rust-back will occur, especially in locations subject to high relative humidity. Should rust-back occur, the entire tank should be reblasted.

VII. COATINGS

There are hundreds of different coatings on the market today designed to meet very specific applications. Years ago, many tank suppliers used a coal-tar coating on their USTs. The coal tar was an excellent coating, and was relatively trouble free. The problem with the coal tar was that it took roughly 24 hours to cure before you could move the tank. The long cure time didn't help tank manufacturers as it created a logjam at the coating application stage.

Coating manufacturers eventually introduced a polyurethane product that provided the necessary corrosion protection and reduced the cure time significantly—enabling the fabricator to move the tank within minutes on a hot day, or just a few hours in cooler weather. With the reduction in curing time most fabricators switched from the coal tar to the polyurethane coatings.

While the polyurethane is a very good coating, it is much more susceptible to application errors or defects. Since the coating is a two-part system, the coating actually mixes at or near the spray nozzle. The ratio by volume of a coating manufacturer's proprietary Part A to Part B in the mixture is critical. Many coating defects are directly attributed to the mixing ratio and may cause the urethane to blister if the ratio is off.

Many jacketed and composite tanks use an isophthalic fiberglass-reinforced polyester (FRP) laminate to provide a corrosion barrier. The fiberglass laminate is an excellent coating for corrosion protection. The application of the laminate requires a skilled operator and well-maintained equipment. The operator must ensure that the correct ratio of glass to resin is maintained at all times. A skilled operator will be able to work around attachments and manways and maintain a holiday-free coating. Most fiberglass-clad tanks require a holiday test—a high-voltage examination of the external tank surface that will discover any exposed metal.

All fabricators will use some kind of inhibitor to protect the tank from ultraviolet light while the tank is in storage. Many fabricators will use a color additive in the final phase of coating for added UV protection. Prior to adding the gel coat color, fabricators can visually inspect the coating for flaws.

Many jacketed tank technologies also use FRP as a corrosion barrier. Because the jacketed tank design calls for the FRP to serve as an outer tank, there is always some spacer material between the steel tank and the FRP shell. This eliminates most of the need to blast the tank. Blasting only occurs where the laminate directly makes contact with the steel. This is usually around the fittings and manway.

Construction of the jacket is critical since it is considered a double-wall tank. The tank surface must be relatively smooth and free of abrupt changes that would create stress on the jacket. Many fabricators reinforce the flanged head and weld seams to reduce, or prevent, such stress. Most jacketed tanks are suspended at the heads during application. The heads are the final area of laminate application, and must be thoroughly inspected during the vacuum integrity test for leakage.

Aboveground tanks require exterior paint for protection from climatic conditions in a variety of areas. If the AST will be located in areas near saltwater, deserts, urban pollution or high humidity, a specifier should check with tank manufacturers and coating suppliers about the recommended application of AST paints.

VIII. INSPECTION

The best time to inspect a UL 58 or UL 142 steel tank is just prior to coating application. Once coated, evaluation of the fabrication and welding is often difficult, if not impossible.

By inspecting prior to coating, you have the opportunity to examine all weld profiles to ensure they meet specification. This also enables the removal of excess spatter. Prior to applying the corrosion barrier, it is important to ensure that no sharp edges are present to inhibit the coating process.

Once the coating has been applied, the inspector should determine if the urethane, FRP, or coal-tar epoxy has the minimum dry-film thickness required by the specification on the entire tank. Inspectors can measure dry-film thickness in a number of tank shell locations with a gage. The coating should be free of sags, runs, blisters, or other surface defects. Composite tanks require a 10,000 to 35,000 V holiday test to ensure that the laminate completely protects the steel from corrosive elements. Check with local officials on requirements. It is recommended that the ground wire lead from the holiday tester be connected directly to the tank for maximum detection of coating defects or flaws.

5
UST and AST Fabrication

Randy Smith
Hamilton Welding Company, Columbus, Ohio

Larry O'Shea
Steel Tank Institute, Lake Zurich, Illinois

I. CREATING WHAT THE END USER NEEDS

Tanks must be fabricated to match an end user's needs. It's true that underground storage tanks (USTs) and aboveground storage tanks (ASTs) are largely standardized to work in a wide range of environments, but many unique situations call for a customized design.

Tank design and fabrication can be influenced by economic factors, regulatory requirements, the liquid to be stored, internal pressures, external environmental forces, corrosion protection, leak detection, and welding needs.

II. ECONOMIC FACTORS

Economic considerations continue to play a major role in the final specification that meshes the tank owner's technical needs as balanced with budgetary limits and regulatory requirements.

A. Single Wall

The single-wall design, the oldest and most widely fabricated UST and AST, is primarily popular because of price. As its name implies, the single-wall tank is fabricated by welding together a shell and heads made of a single thickness of steel. The single-wall underground storage tank is primarily used by service stations and private industrial companies for the storage of fuels that power motorized equipment. The single-wall AST can be found most often on farms and in applications that run no risk of contaminating aquifers or waterways. Another, major use is for the storage of heating fuels.

When regulations don't require secondary containment, many tank buyers will typically not pay the extra premium for additional environmental protection.

B. Double Wall

This is the tank-within-a-tank concept, which creates an interstitial space that can be monitored. Fabricating a built-in leak detection system requires the use of more material and more time by craftsmen at the fabricating plant, so the double-wall designs are naturally more expensive than single-wall tanks.

The interstice (the area between the inner and outer tanks, also known as the annulus) enables leak detection without fear that a hazardous substance has been released to the environment. The secondary tank, or outer shell, is designed to contain any liquid that may have leaked from the primary tank to contaminate soils or water wells.

With the ever-increasing demand for USTs and ASTs to be environmentally friendly, double-wall tanks have gained in popularity. Areas considered ecologically sensitive by regulators are requiring tanks to have secondary containment. Sensitive areas are in close proximity to water supplies—groundwater aquifers, public waterways, lakes, rivers, etc.

Underground tanks storing chemicals or hazardous materials are required by regulation to be double-wall. Chemical storage via an AST can be accomplished with a double-wall tank (or one of several secondary containment options). Many of the larger metropolitan areas are requiring secondary containment, which often means specifiers are asking for the double-wall design.

A steel double-wall underground tank can be fabricated by one of four primary methods:

A tight-wrap design in which a second layer of steel is in intimate contact with the primary tank, which creates a small, but viable, interstitial space

A steel outer tank surrounding a steel inner tank with 2-in. to 3-in. standoffs separating the internal tank from the outer shell

A fiber-reinforced plastic (FRP) outer shell surrounding an inner tank of steel

A high-density polyethylene (HDPE) or polyolefin jacket covering the steel primary tank

A steel double-wall aboveground storage tank is fabricated using the intimate-contact method.

Though tank system costs typically drive the double-wall tank market, there are some special applications for which customers claim that price is no object.

C. Compartment Tanks

The use of compartmentalized tanks is increasing with every year. The ability to store multiple products and save on overall tank system costs has led to increased demand.

This approach allows tanks to be divided into two or more compartments. These tanks are fabricated by adding as many bulkheads as needed to create distinct compartments. Single-wall or double-wall tanks can be fabricated into compartment tanks. A distinct economic advantage is obtained with the use of a secondary containment tank featuring multiple compartments—the end user gets the benefit of several storage containers, but only pays for one secondary containment tank, one interstitial monitoring port and one release detection device. Whether the tank design is single-wall or double-wall, tank owners have reduced costs with compartmentalized tanks by (1) lowering the total number of tanks at one site, which decreases the resources allocated to leak detection, or (2) installing one tank rather than several, which lowers installation costs.

In some states, a compartmentalized tank can save on underground storage tank insurance costs—again, because the owner is insuring one tank rather than several. However, other states treat a three-compartment tank as equivalent to three normal tanks for UST financial responsibility requirements.

With the increasing demand for compartment tanks, the specifier should consider whether fabrication with a single- or double-bulkhead configuration is most appropriate. Bulkheads are basically made the same way as the tank heads. By minimizing the use of weld seams within the bulkhead construction, the risk of leakage from one compartment into another is reduced. Some manufacturers routinely fabricate two- or three-piece bulkheads, which is permitted by Underwriters Laboratories (UL) standards for double-bulkhead designs.

Several methods can assure proper bulkhead construction under the UL 58 (Steel Underground Tanks for Flammable and Combustible Liquids) standard, and model fire codes. The most conservative approach requires two separate bulkheads in a secondary containment design. Under this design, if liquid escaped past one bulkhead, it would drain into an interstitial area. However, a failure at the weld seam between the tank shell and the bulkhead—in the single-bulkhead design—would lead to petroleum leaking directly into the soil (from a single-wall tank) (Fig. 1).

The complete separation of products is a critical factor for compartment tanks that store, for example, gasoline in one compartment and kerosene in another. Commingling of these two flammable and combustible liquids must be pre-

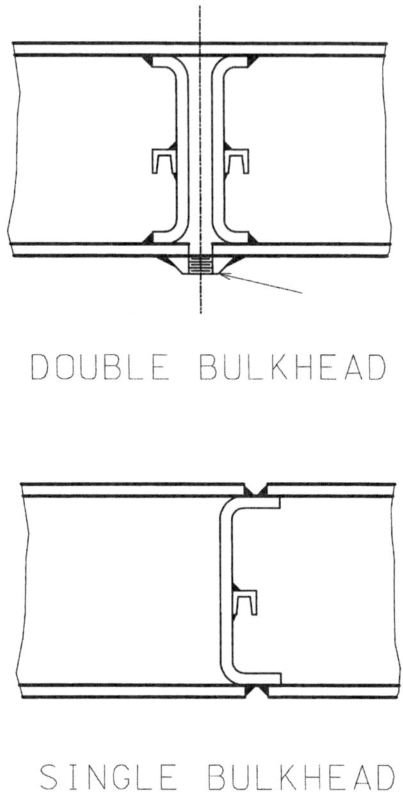

Figure 1 Bulkheads for compartment tanks. Flange for testing may be located anywhere on tank circumference, between two bulkheads.

vented. A camper lighting a stove that accidentally is carrying a mixture of kerosene and gasoline would be subject to an extreme safety hazard.

III. REGULATORY REQUIREMENTS

As mentioned previously, some authorities having jurisdiction (AHJ) may require secondary containment. If so, a double-wall steel or a jacketed tank may fit that underground storage application.

With a double-wall steel tank, communication of a leak into the annular space is not much of an issue for detection. However, the specifier may consider requiring an additional fitting on the annular space—at the tank end opposite of the monitoring port—for removal of fluids in the event of a primary-tank leak. All too often, horror stories from the field are told of deliveries made into an interstitial space, rather than the primary tank. The placement of the additional fitting is not a standard practice among tank manufacturers. This fitting allows easier venting of the interstice, which enables the tank to be reused or retired.

On jacketed tanks, communication of leaks is not so cut and dry. In the UL 1746 (External Corrosion Protection Systems for Underground Storage Tanks) standard—which includes fabrication of jacketed USTs—an interstitial communication test is required on the tank assembly. This is not mandated for double-wall steel.

Why isn't the test required for both? It relates directly to properties of the outer-tank material. For instance, it is impossible to pull two walls of steel together so tightly that an interstice will not exist. But with jacketed tanks, an interstitial vacuum (or geotechnical forces from surrounding soil) can compress the flexible jacket so tightly against the steel primary tank that interstitial communication could be partially eliminated.

IV. LIQUID PRODUCT

Beyond basic sizes and dimensions for a UST or AST, a buyer should consider what would be stored in the tank. For instance, if the product must be maintained at a constant elevated temperature, some tanks (and accessories such as nylon bushings, zinc anodes, and certain coatings) may not be suitable for this application. If the product stored is caustic, the fabricator may need to assemble a stainless steel storage tank.

V. INTERNAL PRESSURES

Fabrication standards for USTs call for building the tanks to atmospheric pressures, capable of withstanding integrity testing of up to 5 psi. The internal pressures on USTs that hold flammable and combustible liquids are seldom a concern because tank burial virtually assures a lack of extreme fluctuations in product temperatures.

Fabricators, however, face an entirely different situation with ASTs loaded with fuel and exposed to sunlight in hot weather, or potentially at risk to catch fire at the tank site. Fire codes require the addition of a fitting for an emergency relief vent to prevent AST fueling tanks from building up explosive internal pressures.

Internal pressures also affect the fabrication of vertical ASTs. Because liquids exert greater pressure at the bottom of a container, vertical aboveground tanks employ thicker-gage steel at the bottom of the cylinder than the top. Vertical ASTs also are tested to hold 2.5 psi vs. 5 psi for horizontal tanks.

VI. EXTERNAL PRESSURES

Soils—and UST backfill materials—shift periodically. Earthquakes and high-water tables created by localized flooding are probably the most dramatic examples of natural events that can place undue pressure on tanks. But years of experience have shown fabricators that special burial conditions can also influence the selection of materials for tank construction. For example, if the tank is going to be buried deeper than 5 ft below grade, additional steel thickness may be required to assure structural integrity.

On rare occasions, deep tank burials take place in areas where high groundwater levels exist. There are documented instances when marginal steel thicknesses have been used in the production of the tank, which deformed the tank bottoms—in essence, crimping the shell upwards. When fabricators are apprised of unusual conditions at the tank site, a UST can be stiffened through thicker walls, or by incorporating steel structural elements. For example, all nonmetallic USTs use reinforcing ribs to obtain the necessary integrity to prevent buckling. When coupled with proper backfill and placement, the FRP tank design will minimize the risk of cracking.

On ASTs, fabricators routinely reinforce larger-diameter tank heads to assure minimal deflection caused by static forces inherent with stored liquids. With underground tanks, the backfill provides resistance to such movement.

VII. CORROSION PROTECTION

Underground storage tanks require corrosion protection to comply with the U.S. Environmental Protection Agency's technical requirements for USTs. On steel tanks, fabricators provide corrosion protection to the primary tank through dielectric coatings, cathodic protection, secondary containment or sometimes a combination of methods (e.g., fiberglass-reinforced plastic coating supplemented by anodes).

Fabricators provide protection against corrosion for ASTs by painting exterior steel surfaces that could be affected by rain, snow, or other climatic conditions. Some ASTs require cathodic protection of the tank bottom when the bottom is in direct contact with corrosive soil, though this is primarily an issue for large-capacity, field-erected tanks.

VIII. LEAK DETECTION EQUIPMENT

The engineer or specifier should consider the type of leak detection equipment to be attached to the tank. Fabricators can provide either internally or externally mounted monitoring pipes for double-wall tanks. An internally mounted interstitial (or secondary containment) monitoring pipe makes coating application easier for the manufacturer, and is standard on many jacketed tank systems.

For ASTs or USTs of small capacity (2000 gal or less), an externally mounted monitoring pipe can be effective when several fittings limit space on the top of the tank. Some tank fabricators will recommend extending the secondary containment steel tank head to provide a monitoring well.

Much of the tremendous growth in demand for small-capacity AST systems relates directly to the ease of monitoring aboveground leaks with a visual inspection. The remediation of an AST system leak also is considerably less expensive than a release from an underground tank. For instance, the AST cleanup requires no tank excavation costs.

IX. WELD JOINTS

Usually the tank manufacturer will determine the weld joint design(s) to be used in fabrication. The engineer or specifier may want to inquire on the type of joints typically employed by the manufacturer, especially on tanks that will have a lining, such as a vessel used for jet fuel storage. Some weld joints are better than others for internally lined tanks.

Some fabricators of vertical ASTs employ the option of a weak roof-to-shell joint to meet venting requirements. If internal forces build to an unacceptably high level, the roof will rise in a hinged motion and release the excess pressure.

6
Storage Tank Specification Considerations

Wayne Stellmach
Rolling Meadows, Illinois

I. DEVELOPING A SPECIFICATION

A storage tank specification can take various forms. It may be prescriptive or performance based. It may address the entire tank system including dispensers and every imaginable bell and whistle or simply cover the tank and basic components such as vents, spill/overfill equipment and leak detection. The tank specification may itself constitute the entire document or it can be but a small component of an extensive project design-and-build.

The customer generally will dictate the scope—or how extensive a specification will be. However, it is the specifying engineer who will likely determine whether a specification is prescriptive or performance-based. Done properly, either approach can lead to a good specification. The difference is that while prescriptive documents enumerate every component down to the specific manufacturer and part number, a performance-based specification incorporates validated third-party standards and recommended practices in establishing basic performance and safety levels.

The principal argument for a prescriptive specification is that by tending to every detail, there can be little room for error. However, performance-based advocates maintain that referencing third-party standards that have already been exhaustively researched provides a solid basis and avoids reinventing the wheel, and potentially erring along the way.

II. PRESCRIPTIVE OR PERFORMANCE-BASED

To illustrate the difference, compare the following specification language for an automatic tank gauge to be used on an underground storage tank (UST) system.

A. Prescriptive

Provide an integrated electronic tank gauging and monitoring system for release detection and inventory control meeting the following requirements: (1) The automatic product level monitor test must detect a 0.2 gal/hr leak rate from any portion of the tank that routinely contains product. (2) Inventory control must detect a release of at least 1.0 percent of flow-through plus 130 gal on a monthly basis.

Tank monitoring system must incorporate magnetostrictive probes in each tank and a monitor with LCD display, printer, audible and visual alarms, and battery backup. System must be UL listed. System shall be Auto-Stik II, as manufactured by EBW Inc., or approved equal.

B. Performance Based

An automatic tank gaging system shall be used for release detection and inventory control and meet the Code of Federal Regulations, 40 CFR Part 280.43 and 280.44. System shall be UL listed and installed in accordance with PEI RP 100.

III. STANDARDIZING A FORMAT

In addition to determining whether the approach is prescriptive or performance based, specifying engineers must also address the specification format. An engineering firm or corporate engineering department should establish a master specification format that all specifiers will follow. This assures not only uniformity and clarity but also that all critical elements of a design specification are addressed.

There are tools available to guide the specifier in the creation of a comprehensive specification. CSI provides templates and software that can be used to create a general specification format. For example, the Steel Tank Institute (STI), as one of several organizations in the petroleum equipment industry providing tank-related software, utilizes the CSI MasterFormat. The Internet is another resource for specifiers with sites ranging from code and standards developers to manufacturers of tanks and components. Many of the equipment manufacturer websites will include downloadable sample specifications. (See Appendix for a partial list of websites with information related to tank codes and standards.)

IV. CITING THIRD-PARTY REQUIREMENTS

Regulations, codes, and standards are, of course, an extremely important area for specifiers to address in a tank system specification. A predominant regulation for UST systems containing petroleum or certain hazardous substances is 40 CFR Part 280 as established by the U.S. Environmental Protection Agency. However, specifiers must also check for state or local regulations or fire codes relating to USTs or aboveground storage tanks (ASTs). Fuel dispensing, in particular, is addressed by fire codes—as are AST installations—and engineers should not even begin to specify such projects before checking with the local fire marshal on what systems and components are allowed. For example, even if a community generally follows a national model fire code such as NFPA 30 or 30A, which allows ASTs for fuel dispensing, the local fire marshal may prohibit such a use for aboveground storage systems.

Several organizations publish standards and recommended installation practices that are routinely referenced in storage tank specifications. In the United States, the most-referenced standards will likely be those of Underwriters Laboratories (UL); in Canada it will be Underwriters Laboratories of Canada (ULC). Many other countries also recognize UL standards. (Details on UL and ULC standards are discussed in Chapters 9, 10, 18, and 19.)

The Petroleum Equipment Institute (PEI) publishes three widely referenced recommended practices (RPs) pertinent to USTs and ASTs:

RP100 (Recommended Practices for Installation of Underground Liquid Storage Systems)

RP200 (Recommended Practices for Installation of Aboveground Storage Systems for Motor Vehicle Fueling)

RP300 (Recommended Practices for Installation and Testing of Vapor Recovery Systems at Vehicle Fueling Sites)

Several other organizations have developed informational resources that may need to be cited in a storage tank system specification:

American National Standards Institute (ANSI)—publishes standards on piping

American Petroleum Institute (API)—publishes RPs for corrosion protection, tank installation, tank entry and lining, closure of tank systems, and storage and handling procedures

American Society for Testing and Materials (ASTM)—publishes information on release detection devices, steel pipe specifications and glass-fiber reinforced polyester USTs

Canadian Petroleum Products Institute (CPPI)—publishes information on UST systems, inventory control, and cathodic protection

NACE International—publishes RPs for corrosion control of tanks and piping
Steel Tank Institute (STI)—publishes RPs for corrosion protection of storage tank systems, tank installation procedures, and tank fabrication standards

By studying codes, regulations, standards, and recommended practices relative to storage tank systems, the design engineer will have a thorough understanding of all the applicable elements of a comprehensive tank system specification such as corrosion protection, spill and overfill equipment, ventilation, secondary containment, leak detection, product dispensing, and safe installation practices.

V. OTHER OPERATIONAL ISSUES

There are still other issues that must be considered, including the applicability of regulations addressing air emissions or spill prevention and response planning. Also important is compatibility of the product with the tank, piping, seals, and other components that come in contact with the product. This is particularly pertinent to chemical storage, where an internal tank liner and special pumps and valves may be necessary.

Specifying engineers should also discuss various practical issues with the customer that will impact operational efficiencies and/or economic/liability concerns. For example, even if secondary containment is not mandated by federal or local regulations, the extra measure of protection it provides merits consideration. Can the customer use compartmented tanks to store multiple products with less tankage and subsequent lower installation and insurance costs? If an aboveground tank is being considered, is there enough real estate at the site to allow for required separation distances from buildings, public rights-of-way, dispensers, and other tanks? Do the potential up-front cost savings for an AST installation outweigh long-term maintenance costs? (For more AST vs. UST considerations, see Chapter 7.)

In short, a "cookie cutter approach" in designing a storage tank system is ill-advised, if not outright dangerous.

VI. SAMPLE SPECIFICATIONS

What follows are two samples of tank project specifications, one for an aboveground storage tank project and one for an underground storage tank project. Compiled by the Steel Tank Institute, these are not intended to be all-inclusive or universally applicable. Nonetheless, they provide good examples of a fairly comprehensive specification. Although these specifications, generated by STI's QuickSpec software, naturally include references to STI storage tank technology, they are provided here to illustrate a "total project perspective" and the vast range of decisions that a specifier has to confront.

These are written in CSI MasterFormat and are designed to allow the specifier latitude to write either a prescriptive or a performance-based specification. Note the bracketed [ ] text, which indicates sections requiring further action or decision on the part of the specifier. To allow relative comparison of an AST to a UST project specification, both are written for a vehicle fueling application.

A. UST System Specification
Division 13—Special Construction
Section 13200—Liquid and Gas Storage Tank Systems
Part 1—General

1. Scope of Section

a. This section describes requirements for providing the equipment, labor, and materials necessary to furnish and install petroleum storage tank system(s) utilizing underground sti-P3 double-wall tank(s).

b. Requirements include furnishing and installing all equipment and accessories necessary to make complete systems for the storage and dispensing of gasoline.

c. The following components shall be provided by the owner and installed by the contractor. [List if applicable.]

d. The following components shall be provided by the contractor, but not be installed as a part of this contract. [List if applicable.]

2. Related Sections

a. All material and installation sections relating to site preparation, painting, concrete, and other related work not specified herein are covered in the appropriate sections.

b. [List section titles and numbers.]

3. Definitions

a. *Agreement* consists of the conditions of the contract between the owner and the contractor, including referenced specifications, drawings, and related documents.

b. *Construction* documents consist of the general and supplemental conditions, specifications, drawings, and any addenda issued prior to bidding.

c. *Contractor* is the person, firm, or corporation with whom owner has entered into the agreement.

d. *Furnish* means the contractor shall supply the item specified, at the job site, unloaded, and secured against damage, vandalism, or theft.

e. *FRP* is an abbreviation for fiberglass-reinforced plastic.

f. *Install* means the contractor shall perform all work required to place the equipment specified in operation, including installation, testing, calibration, and start-up.

g. *Interstitial* refers to any space between primary and secondary containment of tanks as well as containment sumps and piping.

h. *Leak mode testing* refers to testing the integrity of the tanks and in accordance with the test device manufacturer's instructions and U.S. EPA technical standards.

i. *Liquid tight* means prevention of the infiltration of ground or surface water into a contained space, or the release of product from contained spaces into the surrounding soil.

j. *Owner* is the person or entity identified as such in the agreement.

k. *Product* means the gasoline stored and dispensed from the tank.

l. *Provide* means the contractor shall furnish and install the equipment specified, and perform all work necessary to provide a complete and functional system.

m. *Spoil* means all material removed by demolition or excavating.

n. *STI* is the Steel Tank Institute, 570 Oakwood Rd., Lake Zurich, IL 60047, (847) 438-8265.

o. *Substantial completion* is the stage in the progress of the work when the work or designated portion thereof is sufficiently completed in accordance with the contract documents so the owner can utilize the work for its intended use.

p. *Work* means all materials, equipment, construction and services required by the contract, whether completed or partially completed.

4. General Requirements

a. Unless otherwise specified, equipment furnished under this section shall be fabricated and installed in compliance with the instructions of the manufacturer.

b. The contractor shall ensure that all equipment, accessories and installation materials comply with the specification and that adequate provision is made in the tank design and fabrication for mounting the specified system equipment and accessories.

c. The contractor is solely responsible for construction means, methods, techniques, sequences and procedures and for safety precautions and programs.

d. All electrical work shall conform with the National Electrical Code, NFPA 70.

e. The contractor shall provide all labor, equipment and material required to provide a complete and functional system.

f. To avoid delays in construction, the contractor shall ensure that all components of the system are available at the time of installation.

g. The contractor shall coordinate his work with other work being performed at the construction site and to minimize interference with the owner's normal activities which may continue during construction.

h. The contractor shall obtain necessary permits, arrange for inspections and obtain approval of the appropriate authority having jurisdiction over the work described.

5. Standards

a. Work shall be performed in accordance with applicable federal, state, and local fire protection, environmental and safety codes and regulations, and the latest version of the following industry standards:

1. Recommended Practices for Installation of Underground Liquid Storage Systems, PEI/RP100; Recommended Practices for Installation and Testing of Vapor Recovery Systems at Vehicle Fueling Sites, PEI/RP300; Petroleum Equipment Institute, P.O. Box 2380, Tulsa, OK 74101.
2. Installation of Underground Petroleum Storage Systems, API/1615; Cathodic Protection of Underground Petroleum Storage Tank and Piping Systems, API 1632; American Petroleum Institute, 1220 L Street, Washington, DC 20005.
3. Flammable and Combustible Liquids Code, NFPA 30; Automotive and Marine Service Station Code, NFPA 30A; National Electrical Code, NFPA/70; and Underground Leakage of Flammable and Combustible Liquids, NFPA/329; National Fire Protection Association, 1 Batterymarch Park, P.O. Box 9101, Quincy, MA 02269-9904.
4. Article 79—Flammable and Combustible Liquids, Uniform Fire Code, 1994 Edition; International Fire Code Institute, 5360 Workman Mill Rd., Whittier, CA 90601, (310) 699-0124.

5. Hazardous Waste Operations and Emergency Response and Excavating, OSHA/29 CFR 1910.120 & 29 CFR 1926 Subpart P; Occupational Safety and Health Administration, U.S. Department of Labor, Region V, 230 S. Dearborn Street, Room 3244, Chicago, IL 60604.

6. Occupational Safety and Health Standards, Flammable and Combustible Liquids, 29 CFR 1910.106; Personal Protective Equipment 29 CFR 1910 Subpart I, Excavations 29 CFR 1926.650 Subpart P; U.S. Department of Labor, Occupational Safety and Health Administration (OSHA), Washington, DC.

7. Control of External Corrosion of Metallic Buried, Partially Buried, and Submerged Liquid Storage Systems, NACE Recommended Practice RP0285-95; Control of External Corrosion on Submerged Metallic Piping Systems, NACE Recommended Practice RP0169-92; NACE International, P.O. Box 218340, Houston, TX 77213.

8. Installation Instructions, sti-P3, R821; Steel Tank Institute Recommended Practices for Corrosion Protection of Underground Piping Networks Associated with Liquid Storage and Dispensing Systems, R892; Steel Tank Institute Standard for Dual Wall Underground Steel Storage Tanks, F841; Steel Tank Institute, 570 Oakwood Rd., Lake Zurich, IL 60047, (847) 438-8265.

9. Steel Underground Tanks for Flammable and Combustible Liquids, UL Standard 58, 1996; Control Equipment for Use With Flammable Liquid Dispensing Devices, UL 1238; Pipe Connectors for Flammable and Combustible Liquids and LP-Gas, UL 567; Pipe Unions for Flammable and Combustible Liquids and LP-Gas, UL 567; Powered-operated Dispensing Devices for Petroleum Products, UL 87; Valves for Flammable Fluids, UL 842; Corrosion Protection for Underground Storage Tanks, UL 1746; UL Listed Non-Metal Pipe, UL 971; Underwriters Laboratories Inc., 333 Pfingsten Rd., Northbrook, IL 60062, (847) 272-8800.

10. Underground Storage Tanks, Technical Requirements and State Program Approval, Final Rules, 40 CFR Parts 280 and 281, Part II, Federal Register, Friday, Sept. 23, 1988; Musts for USTs: A Summary of the New Regulations for Underground Storage Tank Systems; and Hazardous Waste Management Standards, Federal Register July 14, 1986; U.S. Environmental Protection Agency, Office of Underground Storage Tanks, 401 M Street, SW, Washington, DC 20460.

b. Where differences exist between standards, the contractor shall use the most conservative. If in doubt, describe differences in writing to the owner for his approval before performing the work.

c. The codes and standards listed are the latest as of this publication. Codes and standards are continuously updated. The contractor shall confirm the construction standard edition enforced by the authority having jurisdiction.

6. Submittals

a. The contractor shall provide three (3) sets of shop drawings of the following system components for approval before commencing construction:

1. Shop drawings of the tank(s) by the tank manufacturer
2. Assembly and installation drawings
3. Other [List]

b. The contractor shall provide product data sheets and descriptive material for major components to be provided:

1. Tank(s)
2. Pumps, valves, and fittings
3. Other system accessories [List]

c. Submittals shall be delivered to the engineers within [10 days] of notice to proceed. The engineer shall review the drawings and return them to the contractor approved, or with appropriate comments, within [14 days] of receipt.

7. Guarantees, Warranties, and Insurance

a. The contractor shall provide the following insurance [List type and limits].

b. The contractor shall provide the following guarantees/warranties [List requirements].

8. Documentation

a. The contractor shall provide three (3) sets of the following installation instructions with the submittal of shop drawings:

1. Tank(s)
2. Pumps, valves, pipe and fittings
3. Monitoring system
4. Vapor recovery components
5. Other [List]

Part 2—Products

1. Underground Storage Tank(s)

a. Provide double-wall sti-P3 [Type I] [Type II] underground storage tank(s) for the storage of [petroleum products] at near-atmospheric pressure. Number and size(s) of tank(s) shall be as follows (exact dimensions vary between manufacturers; verify dimensions with manufacturer): 10,000-gal capacity (nominal) tank for gasoline storage. Dimensions to be (diameter, length): 96 in., 27 ft-0 in.

1. The primary and secondary steel tanks shall be manufactured in accordance with UL Standard 58 requirements.
2. Tanks shall have [300] [360] degree double-wall construction.

3. The interstitial space shall be constructed to allow continuous monitoring of the entire interstice for the life of the tank.

4. Metal thickness for tank walls and head shall be per UL 58. Steel shall be equal to ASTM A-36, or better, for chemistry and weldability quality.

5. Manufacturing dimensional tolerances for primary tank shall be −0%/+5%.

6. Tanks shall be protected from corrosion as follows:
   a. Provide a dielectric coating on all exposed metal surfaces.
   b. Provide dielectric bushings or flanged isolation for all tank openings.
   c. Provide galvanic [zinc] [magnesium] anodes [welded] [wired on] to the tank ends.
   d. Provide a [Protection Prover 2 (PP2)] [Protection Prover 1 (PP1) (Note: the PP1 cannot be used with weld-on anodes.)] [equivalent custom-designed] cathodic protection monitor for each non-EPA-regulated tank or tanks of double-wall design. Prior to painting, any connection shall be protected to prevent the coating from later interfering with the integrity of the electrical connection. [d. Provide a Protection Prover 4 (PP4) cathodic protection monitor connection for each tank. Prior to painting, any connection shall be protected to prevent the coating from later interfering with the integrity of the electrical connection.]

7. Tank testing by the tank manufacturer:
   a. The tanks shall be Underwriters Laboratories listed and tested under UL 58.
   b. Tank interior shall be thoroughly cleaned prior to testing.
   c. The tank shall be pressure tested at 3-5 psig and all surfaces soaped and carefully inspected for leaks.

8. Field corrections, repairs, and testing shall be performed in accordance with manufacturer instructions by factory-trained personnel.

b. Tank Requirements

1. The number, sizes, and locations of tank openings are shown in the drawings.
2. Provide 0.25-in.-thick steel striker plates beneath all openings.
3. Number, type, and placement of lifting lugs shall be determined by the manufacturer.
4. A legible placard with the UL label, installation instructions, tank weight, and handling instructions shall be laminated to tank.
5. Tanks shall be compatible with ethanol and methanol fuels.
6. Provide hold-down straps, turnbuckles, and insulating cushioning material to protect the tanks from damage by the straps.

7. Provide deadmen anchors and turnbuckles as required by the tank manufacturer.
8. Exterior tank bottoms shall be protected from physical damage during shipping and storage with protective padding.
9. Acceptable tank manufacturers are [list]:

c. Tanks shall be provided with the following warrantees:

1. 30-year limited warranty against failure caused by non-corrosion related structural failure; corrosion caused by reaction of the tank with its soil environment; and internal corrosion for those tanks equipped with wear plates and used to store heating or motor fuels, including alcohols, and other compatible chemicals.
2. Manufacturer warranty against failure due to defective materials and workmanship for one (1) year following the date of delivery of the tank to the job site.

d. The contractor or the owner shall register each tank and serial number with Steel Tank Institute in accordance with instructions provided by the manufacturer with the tank.

e. Provide a [30] [36] [42] [48] diameter piping containment sump and integral collar with each tank.

1. Riser for [height] burial depth
2. [Fiberglass-coated,] continuously welded, [3⁄16]-in. steel riser
3. The 6-in.-high piping sump collar shall be continuously welded to the primary steel tank, and the pipe sump riser shall be continuously [welded] [bonded] to the collar. All surfaces exposed to the soil shall be cleaned, well coated, and wrapped with fiberglass compatible with the tank material to make a liquid tight seal.
4. Pipe and conduit penetrations fittings shall be liquid tight.
5. Brackets for electrical junction boxes and sensors
6. [24] [30]-in. gasketed, liquid tight manway cover and ring with [threaded stud and wing-nut fasteners] [tie downs] [friction fit] lid

f. Provide a [Protection Prover 1 (PP1)] [Protection Prover 2 (PP2)] [custom-designed] cathodic protection monitor test station for tanks storing non-EPA-regulated products or tanks of double-wall design. [F. Provide a Protection Prover 4 (PP4) cathodic protection monitor test station for tank(s).]

1. The PP4 shall be a preassembled unit consisting of the following:
   a. A buried copper/copper sulfate reference electrode
   b. An at-grade test station with monitoring terminals
   c. Structure leads with connection hardware

2. The PP4 test station shall be designed to monitor one to four tanks on a single test station.
3. The PP4 shall be installed in conjunction with the tank burial by completing the wire connection between the PP4 structure lead and the existing PP2 test lead. Detailed installation instructions shall be provided with each PP4 system, and with sti-P3 Installation Instructions.

2. Fuel-Dispensing Equipment

a. Provide [qty], [qty]-hose, [single] [two] [multi]-product, [product] dispenser(s) with internal Stage II vapor recovery piping and with the following features and accessories:

1. [Electronic] [Mechanical] display of [gallons dispensed] [dollars] [cost per gallon]
2. Internal filters with replaceable cartridges
3. [Painted steel] [Stainless steel] sides, doors, trim and top
4. Nozzles; model [list]
5. Delivery hoses, [length] × [diameter] Coaxial Stage II
6. Hose end swivels; model [list]
7. Hose breakaway valves and connecting hoses; model [list]

[8. Hose retrievers]

b. Acceptable manufacturers are: [list]

c. Provide [qty] [⅓] [¾] [1.5] [2] [3] [5] hp 220 VAC single-phase two-stage submersible turbine pump(s) for [product], model [list]

3. Pump Controls

a. Provide an interface between the liquid sensing system and the pump power which will interrupt power to the submersible pump if the high level liquid sensor located in the containment sump senses the presence of liquid.

1. The system is designed to force recognition of an unacceptably high level of water or released product in contained spaces and to prevent unauthorized restoration of power to the pumps in the event of a shut down.
2. The relays shall be located in the [location].

b. Provide electrical disconnection of all conductors to the suction pump in accordance with NFPA Codes 30, 30A and 70.

1. Locate the emergency shut-off in an accessible area, at least 20 ft but not more than 100 ft from the dispenser. Confirm the final location with the owner prior to installation.
2. Provide a palm type switch button that will shut off electrical power to the pump.
3. The emergency shutoff shall be clearly identified with signage.

4. Emergency shutoff shall have a manual reset.

4. Primary Piping

a. Provide approved [diameter] inch diameter fiberglass underground primary product, Stage II vapor recovery and vent piping.

1. Fiberglass pipe and fittings shall comply with UL 971 standards. All parts shall be manufactured by the same company as part of the same piping system.
2. Use only adhesive provided by pipe manufacturer.
3. Use electric heating collars or chemical heat packs on all joints, regardless of the temperature during construction.
4. Maintain spacing between parallel piping runs of at least twice the product pipe diameter and at least 4-in. spacing between crossed lines.
5. Piping shall be tested in accordance with the manufacturer's instructions and relevant sections of this specification.
6. Provide only female steel pipe thread to male fiberglass pipe thread adapters. Do not join female fiberglass pipe threads to male steel pipe threads.

b. Acceptable manufacturers are: [list]

c. Provide the following riser piping for:

1. [3] [4]-inch galvanized tank fill pipe. Do not coat fill riser with dielectric material. Contact with the soil is essential to dissipate static electricity.
2. [3] [4]-in. Schedule 40 vent [Stage I vapor recovery] riser.
3. 4-in. tank gauge probe.
4. 2-in. interstitial monitor.
5. 2-in. schedule 40 galvanized vent riser aboveground.

d. Provide nonmetallic secondary containment piping for all product piping.

1. Secondary containment piping shall be tested in accordance with the manufacturer's instructions and the Testing section of this Specification.
2. Installation shall be performed by individuals trained by the manufacturer. Contractor shall arrange for on-site training and provide a letter from the manufacturer listing the names of trained individuals and the dates of the training.
3. Acceptable manufacturers are: [list]

5. Monitoring and Gauging System

a. Provide an integrated electronic tank gauging and monitoring system with the following features: [list]

b. Provide liquid sensors in the containment sump(s), programmed to audibly alarm at the low preset level and to interrupt electrical power to the pumps at a high preset level. Acceptable manufacturers are: [list]

c. Provide [magnetostrictive] [mass] [sonic] [infrared] electronic in tank probe(s) capable of detecting a [0.1] [0.2] gal/hr leak. Provide riser with liquid-tight, snap-mounting ring cap kit. Acceptable manufacturers are: [list]

d. Tank manufacturer–provided interstitial liquid sensors shall provide detection of fluid in the interstitial space. Provide all associated conduit and wiring tied into the monitoring system located in the [location], and liquid-tight riser caps. Acceptable monitoring system manufacturers are: [list]

e. Provide [electronic] [mechanical] line pressure leak monitors on product piping capable of detecting a [0.1] [0.2] [0.3] gal/hr leak rate at 10 psi line pressure. Acceptable manufacturers are: [list]

6. Manholes

a. Provide [5] [15] [25]-gallon capacity below grade overfill/spill containment manholes for each fill pipe riser. Acceptable manufacturers are: [list]

b. Provide [size] diameter [bolt-down] [containment] manholes for access to electronic gauge probes and tank interstitial monitor riser. Manhole covers shall be clearly and permanently identified as "Observation Wells." Acceptable manufacturers are: [list]

c. Provide [quantity] [size]-inch triangle, bolt-down observation wells, including plastic-coated steel perforated well casing and clearly-marked lockable lids.

d. Provide 14-in.-diameter bolt-down noncontainment manholes for access to each extractor fittings. Manhole covers shall be clearly and permanently identified as "Observation Wells."

e. Provide [30] [36]-in.-diameter manhole ring and [H-20 rated] [bolt-down] [liquid-tight] [composite] [steel] cover [with recessed handle] for access to [submersible turbine pump] [product piping] sump.

f. Manhole dimensions are nominal. Actual physical pipe sizes, openings, and capacities may vary slightly. Manholes provided must permit free access for use, testing, and repairs to enclosed components.

7. Containment Sump

a. Provide liquid-tight containment sump beneath each dispenser to prevent the release of product into the environment.

b. Sump shall be constructed for use underground and shall be sufficiently reinforced to prevent distortion from the weight of soil, concrete, or groundwater.

c. The containment sump shall be constructed of noncorrodible fiberglass or high-density polyethylene and provided with the following features and accessories:

1. One-inch lip above concrete inside pump perimeter to prevent infiltration of wind-driven rainwater.
2. Internal bracketing for liquid monitoring sensor to eliminate drilling chamber walls.

d. Provide liquid-tight bulkhead fittings for piping and conduit penetrations of the containment sump and gauge containment manhole.

e. Acceptable manufacturers and products are:

1. Containment sumps: [list]
2. Penetration fittings: [list]

8. Steel Island Forms

a. Provide [size (length × width)]-in. high island form with integral pump box designed for the dispenser and incorporating the following features and accessories:

1. [12-gauge] steel island forms
2. [Factory primed. Painted after completion of paving.] [Brushed Stainless Steel]
3. Features to ensure proper alignment

b. Integral dispenser boxes:

1. Provision for anchoring dispenser and piping
2. Provision for securing junction box and conduit

9. Valves, Fittings, Flexible Connectors, and Other Equipment

a. Provide tight, lockable, fill cap and adapter for each tank. Model [list]

b. Provide a [fire impact valve] [shear section] on product pipe beneath each dispenser. Model [list]

c. Provide all-steel flexible connectors at all tank, dispenser, and vent riser connections as shown in the drawings. [Flexible connectors may not be required for system with flexible piping systems]

1. Provide all-steel construction with a UL-listing for use aboveground (UL 567). Do not use connectors with low-melting-point materials.

[2. Flexible connectors for suction piping shall be rated for full vacuum service at 760 mm Mg (mercury) vacuum.]

3. Flexible connectors shall have one swivel end and one female pipe thread end. Units shall be clearly marked with a lay line to minimize chances of twisting during installation.

4. Flexible connectors installed with a 90 degree bend shall be not less than [24] [30]-in. long.

5. Acceptable manufacturers are: [list]

d. Provide isolation boots for each flexible connector in contact with the soil.

1. Isolation boots shall completely isolate the metallic flexible connectors from the soil.
2. A liquid-tight seal which can be tested at not less than 10 psig.
3. Seal to FRP piping with heat shrink material in combination with at least two stainless-steel hose clamps per end. Coat buried clamps with dielectric material after installation.
4. Acceptable manufacturers are: [list]

e. Provide extractable fittings at tank connections as described.

1. Provide each extractor fitting with a ball float valve which extends [6] [12] [16] [24] in. into the tank. The pipe portion shall be equipped with a drain port.

2. Provide brass test plug for each extractor fitting and an extractor wrench.

3. Acceptable manufacturers and model numbers include: [list]

f. Provide [size]-in. upflow [atmospheric] [pressure/vacuum] vent caps. [Pressure vacuum vent caps shall open with [8] oz of pressure and [½] oz of vacuum.] Acceptable manufacturers and models are: [list]

g. Provide overfill valves and [coaxial] [standard] drop tube assemblies for each tank fillpipe. Acceptable manufacturers and models are: [list]

h. Provide portable Class ABC [20 pound] fire extinguisher(s) and weather proof cabinet(s) at dispenser island(s) and [other locations] in accordance with applicable fire codes.

i. Provide [quantity] [size]-diameter [U-shaped steel pipe guards] [bollards] to be placed at the ends of the pump island, primed, and painted.

Part 3—Execution

1. General

a. Familiarity with the Site.

1. Contractor shall familiarize himself with the location of all public utility facilities and structures that may be found in the vicinity of the construction.

2. The contractor shall conduct his operation to avoid damage to the utilities or structures. Should any damage occur due to the contractor's operations, repairs shall be made at the contractor's expense in a manner acceptable to the owner.

3. The contractor is responsible for meeting all the requirements established by the agencies for utility work, as well as work affecting utilities and other government agencies.

2. Excavating and Trenching

a. Excavated materials.

1. Contractor shall remove necessary paving by saw cutting and excavating as required to accomplish the work described on the drawings.

2. Contractor shall temporarily stockpile excavated spoil on site. Contractor shall dispose of clean spoil [on site] [off site].

3. Spoil shall not be considered acceptable as backfill.

3. Backfilling and Compaction

a. Contractor shall provide clean pea gravel, compacted sand or crushed stone backfill for the tank and product piping excavations. All backfill material shall conform with ASTM standard C-33 paragraph 9.1.

1. Pea gravel consisting of naturally rounded particles with a minimum diameter of ⅛ in. and a maximum of ¾ in. as backfill material.

2. Washed crushed stone may be used if it is acceptable to the tank and pipe manufacturers. Crushed stone mix of angular particles with minimum size of ⅛ in. and maximum size of ½ in.

3. Sand shall be clean, well-granulated, free-flowing, noncorrosive, and inert.

4. Provide laboratory analysis (sieve analysis) with preconstruction submittals. All materials must be approved in writing by the engineer prior to placement.

b. Contractor shall carefully place and compact the backfill around the tank, containment sumps, and piping.

1. Take particular care to fully support lower quadrant and sides of tank as well as around the other equipment to prevent future settlement.

2. Sand backfill shall be placed and compacted in 12-in. lifts.

c. If dissimilar materials are used as backfill, or if the surrounding soil is unstable, separate the pea gravel or crushed stone backfill from the surrounding soil through use of geotextile filter fabric.

1. Fabric strips must overlap by at least 18-in, minimum thickness of 4 oz/yd^2.

2. Acceptable filter fabric materials are [list].

4. Tank Handling and Storage

a. Tanks shall be handled, lifted, stored and secured in accordance with the manufacturer's instructions.

b. Unload with equipment having sufficient lifting capacity to avoid damage to the tank. Securely store the tank at the job site.

5. Equipment Installation

a. Contractor shall install tank(s), dispenser(s), piping and equipment in accordance with the manufacturers' installation instructions, industry standard recommended practices and federal, state, and local regulations.

b. Calibration and start-up of equipment shall be performed by factory-trained and qualified personnel.

c. Pipe installation.

1. Maintain at least ⅛ inch slope in all pipe back to the tanks to prevent traps as prescribed in the standards.
2. Provide at least two secondary pipe diameter clearance between parallel piping runs. Separate crossed piping by at least 4 in.

d. The contractor shall provide sufficient tank burial depth to ensure proper piping slope.

1. Tank burial depth shall be a minimum [3′0″] [3′6″] from finished grade, field verify.
2. Field changes or modifications must be approved by owner beforehand.
3. Confirm the tank burial depth with the engineer before excavating.

e. Set steel island form level at 6 in. above general grade. Confirm the placement before placing concrete in the forms.

f. Containment sump.

1. Install containment sump in strict accordance with the instructions of the manufacturer.
2. Support sump securely until it is evenly backfilled with pea gravel.
3. Carefully install penetration fittings for piping and conduit as prescribed by the sump and fitting manufacturers.
4. Sumps which are distorted or not liquid tight after installation shall be replaced at the contractor's expense.

6. Electrical

a. Installation of all electrical components including [list]:

b. Installation shall be in accordance with manufacturers' installation instructions and shall conform to state and local electrical codes with special attention to compliance with requirements for work in classified areas.

c. Provide explosion-proof electrical junction boxes, conduit and sealoffs specified in Article 500-514 of the National Electrical Code.

d. Contractor shall provide wiring and seal-offs for all conduit.

7. Concrete

a. Prior to setting the tanks in place, provide a grading plan showing final elevations and proposed location of control and expansion joints as a shop drawing.

b. Contractor shall provide [6] [8] [10]-in. reinforced concrete paving for [tank] [fuel island area] pad as shown in the drawings.

1. [Provide 6 × 6 #6 WWF reinforcement] [Fibre mesh reinforcement].
2. Miter all corners of pad.

c. Provide an air entrained concrete mix design including plasticizer which will provide [3000] [4000]-psi strength after 28 days, based on Department of Transportation standards for aggregate, mixing, testing, hardness, etc. Concrete mix to be approved by the owner prior to paving.

d. Frame out all street box openings during the general paving.

1. Cast in all street boxes separately to ensure that the lids are set 1-inch above the general finished grade to inhibit infiltration of surface water.
2. Provide dowels at joints of all street boxes as shown on plans.
3. Provide diagonal reinforcement around all manholes as indicated on the drawings.

e. The reinforced concrete pad around tank fill shall be thickened to [8] [10] in. for a distance of twice the opening size to permit the concrete to slope away from the opening to minimize water infiltration.

f. Provide concrete-filled steel island form. Provide at least 6-in. minimum concrete thickness inside island form.

g. Provide compacted backfill and concrete to match existing grades and surfaces.

h. Use of additives for cold weather work must described in the mix design and subject to the approval of the owner.

8. Asphaltic Paving

a. Asphalt pavement shall be replaced to a full depth of [4] [6] [8] in.

b. The existing asphalt pavement shall be sawcut for the entire depth and length of the new construction where full-depth asphalt pavement is to be placed.

c. Materials

1. Full depth asphalt material shall consist of two courses of [2] [3] [4]-in. each for a total depth of [4] [6] [8] in.

2. Bottom course shall be a binder-type material as described in the DOT specifications. This course may be laid in full depth on compacted backfill and mechanically compacted and rolled.

3. Top course shall be a surface-type mix put down and mechanically compacted or rolled in equal separate layers of 2 in. maximum each.

4. All subgrade areas receiving asphalt pavement, full depth or resurfacing, shall be covered with a bituminous material prime coat as specified by DOT.

9. Painting

a. Contractor shall thoroughly clean all manhole trim and covers after the concrete tank pad has set. Contractor shall prime and paint all manhole trim and covers with high gloss exterior enamel. Color shall be in accordance with API 1637.

10. Testing

a. The contractor is responsible for testing all installed systems for liquid tightness and proper operation, including:

1. Preinstallation inspection of all materials.
2. Preinstallation tank tightness testing.
3. Product, containment, and vent piping during construction.
4. Tank risers before backfilling (with inert gas).
5. Containment sump integrity.
6. System tightness test after all work, including paving, is completed and before the system is placed in service.
7. Postinstallation inspection and testing 11 months after substantial completion of all work and approval of the owner.

b. Testing shall be witnessed by the owner.

1. The owner shall witness tank delivery and setting in place, anchoring, backfilling, piping tests, final precision testing and system start-up.

2. The owner shall indicate approval of all testing witnessed in writing.

c. Tests shall be performed in conformance with the manufacturers' instructions, state laws, and the quoted industry standards, particularly PEI/RP100-94.

1. If a conflict exists between the test protocols, the most stringent test shall be performed.

2. Any conflict which affects manufacturers' warranties must be resolved before beginning construction.

3. The contractor shall document all tests in writing, signed by the individuals who performed and witnessed the test.

d. The contractor shall demonstrate the operation of all systems to the owner at the time of the final start-up test.

1. Provide one day of instruction on the proper operation and maintenance of all components.
2. Demonstrations shall include, but are not limited to, pump operation, monitoring and gaging systems, fuel filter replacement, and leak detection.

11. Testing Product and Vent Piping in Accordance with Manufacturer's Instructions and Quoted Industry Standards, Particularly PEI/RP-100-94

a. Maintain minimum 10 psig pressure on all piping during backfilling and paving operations.

12. Testing Containment Sumps

a. After piping and backfilling are completed, the contractor shall perform a hydrostatic test on the containment sumps and manholes as follows:

1. Fill the containment sump with water to a level near the top of the sump and above the penetration fittings. Mark liquid level on the sump wall.
2. The test duration shall be at least 24 hr. Verify that the liquid level has not changed.

b. If the water level changes, find and repair the leaks and repeat the test.

c. After the test, remove all liquids, clean and dry the sumps.

d. Monitor the sumps for liquid infiltration throughout completion of the remaining construction.

13. Test Documentation and Reporting

a. The contractor shall document all testing and provide copies to the owner and authorities having jurisdiction. Test records shall include:

1. Date and time of test
2. Name of tester
3. Names of any inspectors present
4. Test procedure followed
5. Test results

b. Provide documentation for all testing with contract closeout documentation to the owner.

c. The contractor shall ensure that future testing is not impaired. The contractor may be requested to demonstrate the tests as a part of the final approval process.

1. Inspection of tank interstices.

2. Periodic hydrostatic testing of containment sump. Caution: Piping which has contained flammable or combustible liquids may not be air-tested under any circumstances (NFPA 30) [1].

B. AST System Specification
Division 13
Special Construction Section 13200
Liquid and Gas Storage Tank Systems
Part 1—General

1. Scope of Section

a. This section describes requirements for providing the equipment, labor, and materials necessary to furnish and install petroleum storage tank system(s) utilizing a two-hour fire-rated aboveground steel tank(s) with 110 percent secondary containment.

b. Requirements include furnishing and installing all equipment and accessories necessary to make complete systems for the storage and dispensing of gasoline fuel.

c. The following components shall be provided by the owner and installed by the contractor. [List if applicable.]

d. The following components shall be provided by the contractor, but not be installed as a part of this contract. [List if applicable.]

2. Related Sections

a. All material and installation sections relating to site preparation, painting, concrete, and other related work not specified herein are covered in the appropriate sections of this specification.

b. [List section titles and numbers]

3. Definitions

a. Agreement consists of the conditions of the contract between the owner and the contractor, including referenced specifications, drawings, and related documents.

b. Authority having jurisdiction is the [local fire marshal] [building official] [health department] [electrical inspector] [other] having statutory jurisdiction over the project.

c. *Construction* documents consist of the general and supplemental conditions, specifications, drawings, and any addenda issued prior to bidding.

d. *Contractor* is the person, firm, or corporation with whom owner has entered into the agreement.

e. *Furnish* means the contractor shall supply the item specified, at the job site, unloaded, and secured against damage, vandalism, or theft.

f. *FRP* is an abbreviation for fiberglass-reinforced plastic.

g. *Install* means the contractor shall perform all work required to place the equipment specified in operation, including installation, testing, calibration, and startup.

h. *Interstitial* refers to the space between primary and secondary containment of tanks as well as containment sumps and piping.

i. *Leak mode testing* refers to testing the integrity of the tanks and in accordance with the test device manufacturer's instructions and U.S. EPA technical standards.

j. *Liquid tight* means prevention of the release of product from contained spaces into the surrounding soil or the infiltration of ground or surface water into a contained space.

k. *Owner* is the person or entity identified as such in the agreement.

l. *Product* means the [gasoline] [diesel] [fuel oil] [used oil] [other] stored and dispensed from the tank.

m. *Provide* means the contractor shall Furnish and Install the equipment specified, and perform all work necessary to provide a complete and functional system.

n. *Spoil* means all material removed by demolition or excavating.

o. *STI* is the Steel Tank Institute, 570 Oakwood Rd., Lake Zurich, IL 60047, (847) 438-8265.

p. *Substantial completion* is the stage in the progress of the work when the work or designated portion thereof is sufficiently completed in accordance with the contract documents so the owner can utilize the work for its intended use.

q. *Work* means all materials, equipment, construction and services required by the contract, whether completed or partially completed.

4. General Requirements

a. Unless otherwise specified, equipment furnished under this section shall be fabricated and installed in compliance with the instructions of the manufacturer.

b. The contractor shall ensure that all equipment, accessories and installation materials comply with the specification and that adequate provision is made in the tank design and fabrication for mounting the specified system equipment and accessories.

c. The contractor is solely responsible for construction means, methods, techniques, sequences and procedures and for safety precautions and programs.

d. The contractor shall provide all labor, equipment and material required to provide a complete and functional system.

e. To avoid delays in construction, the contractor shall ensure that all components of the system are available at the time of installation.

f. The contractor shall coordinate his work with other work being performed at the construction site and minimize interference with the owner's normal activities which may continue during construction.

g. The contractor shall obtain necessary permits, arrange for inspections and obtain approval of the appropriate authority having jurisdiction over the work described.

5. Standards

a. The manufacture and installation of aboveground storage tank systems described in this section shall adhere to the following standards and regulatory requirements:

1. Standard for Insulated Secondary Containment-Protected Type Aboveground Storage Tanks, Standard UL 2085; Standard for Steel Aboveground Tanks for Flammable and Combustible Liquids, Standard UL 142; Control Equipment for Use With Flammable Liquid Dispensing Devices, UL 1238; Pipe Connectors for Flammable Combustible Liquids and LP-Gas, UL 567; Powered-Operated Dispensing Devices for Petroleum Products, UL 87; Valves for Flammable Fluids, UL 842; UL Listed Non-Metal Pipe, UL 971; Underwriters Laboratories Inc., 333 Pfingsten Rd., Northbrook, IL 60062, (847) 272-8800.
2. Recommended Practices for Installation of Underground Liquid Storage Systems, PEI/RP100; Recommended Practices for Installation of Aboveground Storage Systems for Motor Vehicle Fueling, PEI/RP 200; Recommended Practices for Installation and Testing of Vapor Recovery Systems at Vehicle Fueling Sites, PEI/RP 300; Petroleum Equipment Institute, P.O. Box 2380, Tulsa, OK 74101.
3. Installation of Underground Petroleum Storage Systems, API 1615, American Petroleum Institute, 1220 L St., Washington, DC 20005.
4. Control of External Corrosion of Submerged Metallic Piping Systems, NACE Recommended Practice RP0169, NACE International, P.O. Box 218340, Houston, TX 77213.

5. Protected Aboveground Tanks For Motor Vehicle Fuel-Dispensing Stations Outside Buildings, Article 79, Uniform Fire Code, 1997, International Fire Code Institute.

6. Standard for Thermally Insulated Aboveground Storage Tanks, F941; Fireguard Installation and Testing Instructions for Thermally Insulated, Lightweight, Double Wall Fireguard Aboveground Storage Tanks, R942; Recommended Practice for Corrosion Protection of Underground Piping Networks Associated with Liquid Storage and Dispensing Systems, R892; Steel Tank Institute, 570 Oakwood Rd., Lake Zurich, IL 60047, (847) 438-8265.

7. Flammable and Combustible Liquids Code, NFPA 30, 1996; Automotive and Marine Service Station Code, NFPA 30A, 1996, National Electrical Code, NFPA 70, 1993; National Fire Protection Association, 1 Batterymarch Park, P.O. Box 9101, Quincy, MA 02269-9904.

8. National Fire Prevention Code, 1994, Building Officials and Code Administrators International, 4051 W. Flossmoor Rd., Country Club Hills, IL 60478.

9. Standard Fire Prevention Code, 1995, Southern Building Code Congress International, 900 Montclair Rd., Birmingham, AL 32513-1206.

10. Occupational Safety and Health Standards, particularly Flammable and Combustible Liquids, 29 CFR 1910.106; Personal Protective Equipment 29 CFR 1910 Subpart I, Excavations 29 CFR 1926.650 Subpart P, U.S. Department of Labor, Occupational Safety and Health Administration (OSHA), Washington, DC.

11. Clean Water Act and Oil Pollution Act of 1990, Spill Prevention, Control and Countermeasures (SPCC) Plans, 40 CFR 112, 113 and 114.

12. Applicable state and local regulations and ordinances.

b. In case of differences among building codes, state laws, local ordinances, utility company regulations, and contract documents, the most stringent shall govern.

c. The codes and standards listed are the latest as of this publication. Codes and standards are continuously updated. The contractor shall confirm the construction standard edition enforced by the authority having jurisdiction.

6. Submittals

a. The contractor shall provide three (3) sets of shop drawings of the following system components for approval before commencing construction.

1. Shop drawings of the tank(s) by the tank manufacturer
2. Assembly and installation drawings
3. Other [List]

b. The contractor shall provide product data sheets and descriptive material for major components to be provided.

1. Tank covering material and coatings
2. Pumps, valves, and fittings
3. Other system accessories [List]

c. Submittals shall be delivered to the engineer within [10 days] of notice to proceed. The engineer shall review the drawings and return them to the contractor approved, or with appropriate comments, within [14 days] of receipt.

7. Construction Documentation

a. At contract close-out, the contractor shall provide three (3) sets of the following installation instructions:

1. Tank(s)
2. Pumps, [dispensers], valves, and fittings
3. Vapor recovery components
4. Other [List]

b. The contractor shall provide three (3) sets of manufacturers' system component operation and maintenance manual instructions.

c. The contractor shall provide and record ("as-built") drawings and photographs of the following:

1. All underground system components
2. The completed tank system in place

d. The contractor shall provide copies of all testing and inspection reports to the owner prior to substantial completion.

8. Guarantees, Warranties, and Insurance

a. The contractor shall provide the following insurance [List type and limits].

b. The contractor shall provide the following guarantees/warranties [List requirements].

Part 2—Products

1. Fire-Rated Aboveground Storage Tank(s) (Tank 1)

a. The storage tank(s) shall be Fireguard aboveground tank(s) for the storage of petroleum product(s) at near atmospheric pressure. Number, size(s) and weight(s) of tank(s) shall be as follows (exact dimensions and weights vary between manufacturers; verify with manufacturer): (1) 10,000-gal capacity (nominal) cylindrical tank for gasoline storage. Dimensions to be (diameter, length): 102 × 330 in.

1. The primary and secondary tanks shall be manufactured in accordance with Steel Tank Institute Publication No. F941, "Standard for Thermally Insulated Aboveground Storage Tanks."

2. The listed assembly shall meet the requirements for "protected" tank as defined by the UFC Article 79 and "fire resistant" tanks as defined by Underwriters Laboratories including impact resistance, ballistics protection and hose stream resistance criteria.

3. The tank shall consist of an inner steel wall, encased by lightweight thermal insulation material, and an outer steel wall.

4. The outer steel wall shall be UL 2085 listed for secondary containment and capable of providing a minimum 110-percent containment of the primary storage tank's content.

5. A legible UL 2085 label shall be affixed to the side of the aboveground storage tank(s).

6. Steel outer wall of the tank shall be coated to prolong weather resistance and to further reduce maintenance needs.

7. The storage tank and supports shall be delivered as a complete UL-listed unit.

8. The storage tank and supports shall meet all the requirements for Seismic Zone 1 per Uniform Building Code requirements.

b. Tank(s) shall be designed for use aboveground and include integral secondary containment, and thermal insulation that provides a minimum two-hour fire rating.

1. Provide a porous, lightweight monolithic thermal insulation material in the tank's interstitial space.

2. The thermal insulating material shall allow liquid to migrate through the interstice to the monitoring point.

3. The thermal insulation material shall not be exposed to weathering and shall be protected by the steel secondary containment outer wall.

4. Thermal insulation material shall be installed at the factory and be in accordance with American Society for Testing and Materials (ASTM) Standards C-332 and C-495.

c. Tanks shall be provided with the following warrantees:

1. 30-year limited warranty against leakage from the secondary containment tank, and failure of the primary tank caused by cracking, breakup, or collapse.

2. 30-year warranty that the tank(s) was fabricated in accordance with requirements of UL 2085 and UL 142, aboveground storage tank manufacturing standards of Underwriters Laboratories.

3. One (1)-year warranty against failure due to defective materials and workmanship for one (1) year following the date of delivery of the tank to the job site.

d. The [contractor] [owner] shall register each tank and serial number with Steel Tank Institute in accordance with instructions provided by the manufacturer with the tank.

e. Tanks with capacities of 5000 gal or more shall be provided with an [18-in.]-diameter manway to allow access into tanks.

f. Provide an interstitial monitoring tube for monitoring the tank's interstice for liquids.

g. Provide an external stairway [safety platform/catwalk] to allow access to top of tank for filling and maintenance that complies with applicable OSHA standards and building codes.

2. Venting Requirements

a. Provide one (1) normal atmospheric or pressure/vacuum vent for the primary tank(s).

1. Vents may discharge upward or laterally, be protected from intrusion of rain, and incorporate a flame arrestor.
2. When applicable, tanks located in Stage II vapor recovery mandated air quality areas shall be provided with pressure/vacuum vents.
3. Vents for tanks containing Class 1 liquids shall terminate at least 12 ft above ground level and be located at least 5 ft from building openings.
4. Vent installation shall comply with applicable sections of the fire and mechanical codes, including, but not limited to, NFPA 30A (2-4.5.e), NFPA 30 (2-3.5), or UFC (7902.1.10), or BOCA (F-3201.1).
5. Accepted manufacturers and part numbers: [List]

b. Provide one (1) emergency primary tank vent per tank.

1. Vent size shall be determined by the tank configuration, the primary tank capacity, and the product stored.
2. Emergency venting shall comply with provisions of NFPA 30A (2-4.5.f), NFPA 30 (2-3.6), or UFC (7902.2.6), or BOCA (F-3201.1).
3. Accepted manufacturers and part numbers: [List].

c. Provide one (1) emergency vent for each secondary containment tank interstice.

1. The venting capacity is determined by the tank configuration, secondary tank capacity, and the product stored.
2. Emergency venting shall comply with provisions of NFPA 30A (2-4.5), NFPA 30 (2-3.6), or UFC (7902.2.6) (Article 79 5.3), or BOCA (F-3201.1), and UL 142, and UL 2085.
3. Vents shall be located as close to the center of the tank as possible.
4. Accepted manufacturers and part numbers: [List].

3. Tank Filling and Overfill Prevention Components

a. For tank(s) with [remote] [side] fill assembly, provide a lockable dry break adapter, lid, or check valve and quick connect coupling for each tank.

1. Protect remote fills from physical damage, tampering and vandalism.

2. Comply with provisions of NFPA 30 (2-4.6), NFPA 30A (2-4.9.4) or UFC Article 79 (5.6).

3. Anti-siphon valve shall be provided for all remote fill or low mounted pump applications, in accordance with NFPA 30A (2-4.6.2 and 2-4.6.3); or UFC Article 79 (6.4); or BOCA (F-3207.5.6). Anti-siphoning may be met by modifying the drop tube, piping or by installation of an electric solenoid valve.

[4. Provide a Stage I vapor recovery system to capture displaced vapors during tank filling process.]

b. Provide a spill container with cover and lockable hasp to contain product spills from the fill hose. Spill containment shall comply with UFC Article 79 (5.7).

1. Fill pipe spill container shall have a capacity of not less than five gallons.

2. Provide a means for returning collected product to the storage tank (not applicable to remote or side fill).

c. Provide overfill prevention equipment which complies with the requirements of NFPA 30A (2-4.6.1) and which incorporates the following features:

1. An audible alarm which will sound when the product level in the tank has reached 90 percent of tank capacity.

[2. A positive shut-off fill limiter which will stop the flow of liquid into the tank when product level reaches 95 percent of tank capacity.]

[3. The limiting device shall be rated to accept the fill flow rate and pressure.]

4. Acceptable Manufacturers: [List].

[Overfill Prevention—Alternative Language if Based on Uniform Fire Code].

c. Provide overfill prevention equipment which complies with the requirements of UFC Article 79 (5.5) and which incorporates the following features:

1. An audible or visual alarm when product level in the tank has reached 85 percent of tank capacity, or a liquid level gauge marked at 85 percent of tank capacity.

[2. A positive shut-off fill limiter which will stop the flow of liquid into the tank when product level reaches 90 percent of tank capacity.]

[3. The fill limiting device shall be rated to accept the fill flow rate and pressure.]

4. Acceptable Manufacturers: [List].

[Overfill Prevention—Alternative Language if Based on the BOCA Code].

d. Provide overfill prevention equipment which complies with the requirements of BOCA (F-3207.5.4) and incorporates the following features:

1. An audible or visual alarm when product level in the tank has reached 85 percent of tank capacity, or a liquid level gauge marked at 85 percent of tank capacity.

[2. A positive shut-off fill limiter which will stop the flow of liquid into the tank when product level reaches 90 percent of tank capacity.]

[3. The fill limiting device shall be rated to accept the fill flow rate and pressure.]

4. Acceptable Manufacturers: [list].

e. The contractor shall provide a means for determining the liquid level in tank which is accessible to the delivery operator, in accordance with NFPA 30A (2-4.6.1) or UFC Article 79 (5.5).

4. Monitoring and Gauging System

a. Provide leak detection for each tank interstice to continuously monitor both the primary and secondary containment tanks.

b. Probe shall be installed in each storage tank's interstitial monitoring tube. The location of the monitoring console and external alarms are noted on the engineering drawings.

c. Provide one (1) electronic tank gauging and monitoring system with the following features:

1. Tank probe(s) for liquid level sensing
2. Control console with display and printer
3. Means of water detection
4. External communications capabilities
5. Tank liquid level and water measurement

d. Provide [electronic] [mechanical] line pressure leak monitors on product piping capable of detecting a [0.1] [0.2] [0.3] gal/hr leak rate at 10 psi line pressure.

e. Acceptable manufacturers and model numbers are: [List].

5. Piping (as Applicable)

a. Aboveground piping shall be Schedule 40 steel pipe with standard (150#) malleable iron fittings.

1. Exposed piping shall be protected from exposure to outdoor conditions.
2. Low melting point materials may be not be used aboveground.

b. Provide approved [diameter] inch diameter fiberglass underground primary product, Stage II vapor recovery and vent piping.

1. Fiberglass pipe and fittings shall comply with UL 971 standards. All parts shall be manufactured by the same company as part of the same piping system.
2. Use only adhesive provided by pipe manufacturer.
3. Use electric heating collars or chemical heat packs on all joints, regardless of the temperature during construction.
4. Maintain spacing between parallel piping runs of at least twice the product pipe diameter and at least 4-in. spacing between crossed lines.
5. Piping shall be tested in accordance with the manufacturer's instructions and relevant sections of this specification.
6. Provide only female steel pipe thread to male fiberglass pipe thread adapters. Do not join female fiberglass pipe threads to male steel pipe threads.

c. Acceptable manufacturers are: [list].

d. Provide nonmetallic secondary containment piping for all product piping.

1. Secondary containment piping shall be tested in accordance with the manufacturer's instructions and the testing section of this specification.
2. Installation shall be performed by individuals trained by the manufacturer. Contractor shall arrange for on-site training and provide a letter from the manufacturer listing the names of trained individuals and the dates of the training.
3. Acceptable manufacturers are: [list].

6. Valves, Fittings, and Flexible Connectors

a. Provide a [fire impact valve] [shear section] on product pipe beneath each dispenser. Model [list].

b. Provide all-steel flexible connectors at all tank, dispenser and vent riser connections as shown in the drawings. [Flexible connectors may not be required for system with flexible piping systems.]

1. Provide all-steel construction with a UL listing for use aboveground (UL 567). Do not use connectors with low melting point materials.
2. Flexible connectors shall have one swivel end and one female pipe thread end. Units shall be clearly marked with a lay line to minimize chances of twisting during installation.
3. Flexible connectors installed with a 90-degree bend shall be not less than [24] [30] in. long.

4. Acceptable manufacturers are: [list].

c. Provide isolation boots for each flexible connector in contact with the soil.

1. Isolation boots shall completely isolate the metallic flexible connectors from the soil.
2. A liquid-tight seal that can be tested at not less than 10 psig.
3. Seal to FRP piping with heat shrink material in combination with at least two stainless-steel hose clamps per end. Coat buried clamps with dielectric material after installation.
4. Acceptable manufacturers are: [list].

d. Provide a steel or nodular iron block valve to allow the tank and piping to be isolated and secured.

e. Provide an anti-siphon device in the product piping at the tank which will prevent the flow of liquid from the tank unless the dispenser is operating, in accordance with NFPA 30A (2-4.6.2 and 2-4.6.3); UFC Article 79 (6.4); BOCA (F-3207.5.6). Anti-siphoning may be met by modifying the drop tube, piping or by installation of an electric solenoid valve. A solenoid valve used in conjunction with an under-pump vacuum-operated valve meets this requirement.

f. Provide a pressure-relief valve in each segment of blocked piping which will relieve excessive pressure resulting from thermal expansion and return any excess product to the tank.

g. Provide portable Class ABC [20 lb] fire extinguisher(s) and weatherproof cabinet(s) at dispenser island(s) and [other locations] in accordance with applicable fire codes.

h. Provide [quantity] [size] diameter [U-shaped steel pipe guards] [bollards] to be placed at the ends of the pump island, primed and painted.

7. Pumping Equipment, Valves, and Fittings—Fuel-Dispensing Applications

a. Provide [qty], [qty]-hose, [single] [two] [multi]-product, [product] dispenser(s) [with internal Stage II vapor recovery piping and] with the following features and accessories:

1. [Electronic] [Mechanical] display of [gallons dispensed] [dollars] [cost per gallon]
2. Internal filters with replaceable cartridges
3. [Painted steel] [Stainless-steel] sides, doors, trim and top
4. Nozzles; model [list]
5. Delivery hoses, [length] × [diameter] coaxial Stage II

6. Hose end swivels; model [list]
7. Hose breakaway valves and connecting hoses; model [list]
[8. Hose retrievers]

b. The dispenser and its components shall be Underwriters Laboratories listed for the purpose intended and shall comply with the requirements of NFPA 30A (4-2.5, 4-2.7), or UFC (5201 and 5202) or BOCA (F-3201.1 and 3207).

c. Acceptable manufacturers are: [list].

d. Provide [qty] [1/3] [3/4] [1.5] [2] [3] [5] hp 220 VAC single-phase two-stage submersible turbine pump(s) for [product], model [list].

e. Provide a listed shut-off valve with a shear section. The valve shall be installed in strict accordance with the manufacturer's instructions.

1. The valve shall be installed directly under dispenser with the shear section at the same level as the top of the island on which the dispenser is located in accordance with NFPA 30A (2-4.6.5), or UFC (5202.5.3.2); or BOCA (F-3205.9).

f. Provide a Stage II vapor recovery system approved for use with aboveground tank systems by the California Air Resources Board, NFPA 30A (4-4 and 4-5) or UFC (5202.12).

8. Pump Controls

a. Provide an interface between the liquid sensing system and the pump power which will interrupt power to the pump if the high level liquid sensor located in the containment sump senses the presence of liquid.

1. The system is designed to force recognition of an unacceptably high level of water or released product in contained spaces and to prevent unauthorized restoration of power to the pumps in the event of a shutdown.
2. The relays shall be located in the [location].

b. Provide electrical disconnection of all conductors to the pump in accordance with NFPA Codes 30, 30A and 70.

1. Locate the emergency shut-off in an accessible area, at least 20 ft but not more than 100 ft from the dispenser. Confirm the final location with the owner prior to installation.
2. Provide a palm type switch button that will shut off electrical power to the pump.
3. The emergency shutoff shall be clearly identified with signage.
4. Emergency shutoff shall have a manual reset.

9. Island Forms

a. If the dispensing pump is located away from tank, it shall be securely anchored to a concrete island and otherwise protected against collision damage in accordance with NFPA 30A (4-2.5) or UFC (5201.5.1).

b. Provide [size (length × width)]-in. high island form with integral pump box designed for the dispenser and incorporating the following features and accessories:

1. [12-gauge] steel island forms.
2. [Factory primed. Painted after completion of paving.] [Brushed Stainless Steel]
3. Features to ensure proper alignment.

10. Containment Sump

a. Provide liquid-tight containment sump beneath each dispenser to prevent the release of product into the environment.

b. Sump shall be constructed for use underground and shall be sufficiently reinforced to prevent distortion from the weight of soil, concrete, or groundwater.

c. The containment sump shall be constructed of noncorrodible fiberglass or high-density polyethylene and provided with the following features and accessories:

1. One-inch lip above concrete inside pump perimeter to prevent infiltration of wind-driven rainwater.
2. Internal bracketing for liquid monitoring sensor to eliminate drilling chamber walls.

d. Provide liquid-tight bulkhead fittings for piping and conduit penetrations of the containment sump and gauge containment manhole.

e. Acceptable manufacturers and products are:

1. Containment sumps: [list]
2. Penetration fittings: [list]

Part 3—Execution

1. General

a. Familiarity with the Site.

1. Contractor shall familiarize himself with the location of all public utility facilities and structures that may be found in the vicinity of the construction.
2. The contractor shall conduct his operation to avoid damage to the utilities or structures. Should any damage occur due to the contractor's operations, repairs shall be made at the contractor's expense in a manner acceptable to the owner.
3. The contractor is responsible for meeting all the requirements established by the agencies for utility work, as well as work affecting utilities and other government agencies.

2. Site Preparation

a. The site shall be prepared to ensure adequate support for the tank system and drainage of surface water.

1. The foundation and tank supports shall be capable of supporting the weight of the tank and associated equipment when full.

2. The foundation may be comprised of concrete, asphalt, gravel or other stable material designed to prevent tank movement, and must be rated for the seismic zone noted in Section II for each tank.

b. Diking may not be required for Fireguard tanks.

1. Regional and local fire code authorities should be consulted for local requirements.

2. Notify the engineer of any local requirements not incorporated in the system as designed.

c. Provide a chain link fence at least 6 ft high, separated from the tanks by at least 10 ft and having a gate that is properly secured against unauthorized entry. (NFPA 30A, 2-4.7.1)

1. Regional and local fire code authorities should be consulted for local requirements.

2. Fencing at the tank area is not required by NFPA if the property on which the tanks are located is secured with a perimeter security fence.

d. Provide barriers around aboveground tanks to protect the tank(s) against vehicular collision in accordance with fire regulations and building codes. NFPA 30A (2-4.7.1) or BOCA (F-3205.4).

e. Maintain legal separation distances from property lines, buildings, public ways, dispensers, vehicles being fueled, and other storage tanks.

1. Caution: Distance requirements vary significantly between jurisdictions.

2. National standards dealing with set back and separation distances are included in NFPA 30A (2-4.2.2), or UFC Article 79 (5.1), or BOCA (F-3207.5.7).

3. Tank Handling, Storage, and Installation

a. Tanks shall be handled, lifted, stored, and secured in accordance with the manufacturer's instructions.

b. Unload with equipment having sufficient lifting capacity to avoid damage to the tank. Securely store the tank at the job site.

c. The tank and associated equipment shall be installed in accordance with the fire safety codes, regulations, standards, and manufacturers' instructions including:

1. Federal, state, and local fire safety, occupational health, and environmental regulations.
2. Steel Tank Institute installation instructions for Fireguard aboveground tanks (Publication No. R942, Installation and Testing Instructions for Thermally Insulated Lightweight Double Wall Fireguard Aboveground Storage Tanks).
3. The installation instructions of other system component manufacturers.
4. The construction documents and associated drawings.
5. Recommended Practices for Installation of Aboveground Storage Systems for Motor Vehicle Fueling, PEI/RP 200, Petroleum Equipment Institute.

d. Advise the owner of any shipping or handling damage encountered.

e. No modifications shall be made to any tank without the prior written approval of the manufacturer and the engineer. This includes any welding on tank shells, adding penetrations in the tank structure, or repairing damage that might affect the integrity of the inner or outer tank.

4. Corrosion Protection

a. Any portion of the fueling system in contact with the soil shall be protected from corrosion in accordance with sound engineering practice and in accordance with NFPA 30A (2-4.8).

b. Protect exposed piping and equipment from corrosion by painting or wrapping it with a coating which is compatible with the product stored and the conditions of the exposure.

5. Excavating and Trenching (for Remote Systems with Buried Piping)

a. Excavated materials.

1. Contractor shall remove necessary paving by saw cutting and excavating as required to accomplish the work described on the drawings.
2. Contractor shall temporarily stockpile excavated spoil on site. Contractor shall dispose of clean spoil [on site] [off site].
3. Spoil shall not be considered acceptable as backfill.

6. Backfilling and Compaction

a. Contractor shall provide clean pea gravel, compacted sand, or crushed stone backfill for the product piping excavations. All backfill material shall conform with ASTM standard C-33 paragraph 9.1.

1. Pea gravel consisting of naturally rounded particles with a minimum diameter of ½ in. and a maximum of ¾ in. as backfill material.

2. Washed crushed stone may be used if it is acceptable to the pipe manufacturers. Crushed stone mix of angular particles with minimum size of ⅛ in. and maximum size of ½ in.

3. Sand shall be clean, well granulated, free flowing, noncorrosive, and inert.

4. Provide laboratory analysis (sieve analysis) with preconstruction submittals. All materials must be approved in writing by the engineer prior to placement.

b. Contractor shall carefully place and compact the backfill around the tank, containment sumps, and piping.

c. If dissimilar materials are used as backfill, or if the surrounding soil is unstable, separate the pea gravel or crushed stone backfill from the surrounding soil through use of geotextile filter fabric.

1. Fabric strips must overlap by at least 18 in., minimum thickness of 4 oz/yd^2.

2. Acceptable filter fabric materials are [list].

7. Equipment Installation

a. Contractor shall install tank(s), dispenser(s), piping and equipment in accordance with the manufacturers' installation instructions, industry standard recommended practices and federal, state, and local regulations.

b. Calibration and start-up of equipment shall be performed by factory-trained and qualified personnel.

c. Pipe installation (as applicable).

1. Maintain at least ⅛ in. slope in all pipe back to the tanks to prevent traps as prescribed in the standards.

2. Provide at least two secondary pipe diameter clearance between parallel piping runs. Separate crossed piping by at least 4 in.

d. Set steel island form level at 6 in. above general grade. Confirm the placement before placing concrete in the forms.

e. Containment sump.

1. Install containment sump in strict accordance with the instructions of the manufacturer.
2. Support sump securely until it is evenly backfilled with pea gravel.
3. Carefully install penetration fittings for piping and conduit as prescribed by the sump and fitting manufacturers.
4. Sumps that are distorted or not liquid tight after installation shall be replaced at the contractor's expense.

8. Electrical

a. Installation of all electrical components including [list].

b. Installation shall be in accordance with manufacturers' installation instructions and shall conform to state and local electrical codes with special attention to compliance with requirements for work in classified areas.

c. Provide explosion-proof electrical junction boxes, conduit and seal-offs specified in Article 500-514 of the National Electrical Code.

d. Contractor shall provide wiring and sealoffs for all conduit.

9. Concrete

a. Prior to setting the tanks in place, provide a grading plan showing final elevations and proposed location of control and expansion joints as a shop drawing.

b. In addition to the seismic-rated concrete foundation footings for the storage tank, contractor shall provide [6] [8] [10]-in. reinforced concrete paving for [tank] [fuel island area] pad as shown in the drawings.

1. [Provide 6 × 6 #6 WWF reinforcement] [Fibre mesh reinforcement].
2. Miter all corners of pad.

c. Provide an air-entrained concrete mix design including plasticizer that will provide [2000] [3000] [4000] psi strength after 28 days, based on Department of Transportation (DOT) standards for aggregate, mixing, testing, hardness, etc. Concrete mix to be approved by the owner prior to paving.

d. Frame out all street box openings during the general paving.

1. Cast in all street boxes separately to ensure that the lids are set 1 in. above the general finished grade to inhibit infiltration of surface water.
2. Provide dowels at joints of all street boxes as shown on plans.
3. Provide diagonal reinforcement around all manholes as indicated on the drawings.

e. Provide concrete-filled steel island form. Provide at least 6-in. minimum concrete thickness inside island form.

f. Provide compacted backfill and concrete to match existing grades and surfaces.

g. Use of additives for cold-weather work must be described in the mix design and subject to the approval of the owner.

10. Asphaltic Paving

a. Asphalt pavement shall be replaced to a full depth of [4] [6] [8] in.

b. The existing asphalt pavement shall be sawcut for the entire depth and length of the new construction where full-depth asphalt pavement is to be placed.

c. Materials

1. Full-depth asphalt material shall consist of two courses of [2] [3] [4] in. each for a total depth of [4] [6] [8] in.

2. Bottom course shall be a binder-type material as described in the DOT specifications. This course may be laid in full depth on compacted backfill and mechanically compacted and rolled.

3. Top course shall be a surface-type mix put down and mechanically compacted or rolled in equal separate layers of 2 in. maximum each.

4. All subgrade areas receiving asphalt pavement, full-depth or resurfacing, shall be covered with a bituminous material prime coat as specified by DOT.

11. Testing

a. The contractor is responsible for testing all installed systems for liquid tightness and proper operation, including:

1. Preinstallation inspection of all materials.
2. Product, containment, and vent piping during construction.
3. Containment sump integrity.
4. System tightness test after all work, including paving, is completed and before the system is placed in service.
5. Postinstallation inspection and testing 11 months after substantial completion of all work and approval of the owner.

b. Test each component of the system for calibration, tightness and proper operation in accordance with the instructions of the component manufacturer.

c. Testing shall be documented by the contractor and witnessed by the Engineer.

1. Record the date and time of the test, the name of the tester and his affiliation with the project, and the names of each individual witnessing the test.
2. Record the test method, duration, and results.
3. Provide a record of the testing to the owner at the time of system startup.

d. Testing shall be witnessed by the owner.

1. The owner shall witness tank delivery and setting in place, anchoring, backfilling, piping tests, final precision testing, and system startup.

2. The owner shall indicate approval of all testing witnessed in writing.

e. Tests shall be performed in conformance with the manufacturers' instructions, state laws, and the quoted industry standards, particularly PEI RP 200 and PEI RP 100.

1. If a conflict exists between the test protocols, the most stringent test shall be performed.
2. Any conflict which affects manufacturers' warranties must be resolved before beginning construction.
3. The contractor shall document all tests in writing, signed by the individuals who performed and witnessed the test.

f. The contractor shall demonstrate the operation of all systems to the owner at the time of the final start-up test.

1. Provide one day of instruction on the proper operation and maintenance of all components.
2. Demonstrations shall include, but are not limited to, pump operation, monitoring, and gaging systems, fuel filter replacement, and leak detection.

12. Testing Primary and Secondary Tanks

a. Air pressure testing of the inner tank and secondary containment tank shall be conducted on-site, in the presence of the engineer, before placing the tank in service.

b. Refer to STI Publication No. R942 for complete procedural details.

c. Other integrity tests may be required by the local authority having jurisdiction.

13. Testing Product, Containment, and Vent Piping

a. Testing product and vent piping in accordance with manufacturer's instructions and quoted industry standards, particularly PEI RP 100.

b. Maintain minimum 10 psig pressure on all piping during backfilling and paving operations.

c. Tightness test the piping at 50 psig before it is buried or connected to the tank. Soap and examine all connections for leakage.

d. Tightness test the secondary containment piping with [5] [10] [20] psig before it is buried or connected to the tank. Soap and examine all connections for leakage.

14. Testing Containment Sumps

a. After piping and backfilling are completed, the contractor shall perform a hydrostatic test on the containment sumps and manholes as follows:

1. Fill the containment sump with water to a level near the top of the sump and above the penetration fittings. Mark liquid level on the sump wall.

2. The test duration shall be at least 24 hr. Verify that the liquid level has not changed.

b. If the water level changes, find and repair the leaks and repeat the test.

c. After the test, remove all liquids, clean and dry the sumps.

d. Monitor the sumps for liquid infiltration throughout completion of the remaining construction.

15. Test Documentation and Reporting

a. The contractor shall document all testing and provide copies to the owner and authorities having jurisdiction. Test records shall include:

1. Date and time of test
2. Name of tester
3. Names of any inspectors present
4. Test procedure followed
5. Test results

b. Provide documentation for all testing with contract close-out documentation to the owner.

c. The contractor shall ensure that future testing is not impaired. The contractor may be requested to demonstrate the tests as a part of the final approval process.

1. Inspection of tank interstices.

2. Periodic hydrostatic testing of containment sump. Caution: Piping which has contained flammable or combustible liquids may not be air tested under any circumstances (NFPA 30) [2].

REFERENCES

1. Steel Tank Institute. sti-P_3 QuickSpec 1.1 CD-ROM. Lake Zurich, IL, 1998.
2. Steel Tank Institute. Fireguard QuickSpec 3.0 CD-ROM. Lake Zurich, IL, 1998.

7

To Bury or Not to Bury: Steel Tank Technology Decisions

Wayne B. Geyer
Steel Tank Institute, Lake Zurich, Illinois

As you explore the handbook's next two sections on underground storage tank (UST) and aboveground storage tank (AST) systems, many questions may surface. The most basic question will be, Which makes more sense for my project—an underground or aboveground system? Let's take a look at some key questions that influence the ultimate specification.

I. UST OR AST?

A. Zoning Requirements and Applicable Regulations Permit ASTs?

The local authority has total control over whether a tank permit will be issued. They review plans to assure compliance with local, state, and federal requirements. If the tank doesn't meet applicable codes, don't expect to get an approval. There are many cases in which government officials determine that the tank system will place undue risks on the general public, even though the system has been designed per national codes. Some give-and-take is often necessary to minimize the risks and satisfy the authority that the permit can be issued. If the answer to question A is yes, go to B.

B. Is Adequate Real Estate Available to Separate the Tank Safely from Traffic, Buildings, Property Lines, and Public Ways—Today and in the Future?

It generally doesn't make sense to spend $100,000 on a tank system today only to be required to spend much more in two or three years simply because you didn't plan properly for a new building or roadway expansion. If the answer to question B is yes, go to D.

If the answer to A or B is no, go to C.

C. Can a Variance from the Authority Having Jurisdiction (AHJ) be Obtained if Adequate Additional Safety Measures are Installed?

Some jurisdictions simply do not allow ASTs for fueling. Or there might be severe size restrictions. The only solution is to apply for a variance. But don't be surprised if additional safety features are required to satisfy the AHJ. For example, one state allows ASTs for fueling up to 1000 gal capacity. However, if a protected tank meeting Underwriters Laboratories UL 2085 standard is installed, larger tank capacities are acceptable. If the answer to question C is no, then UST installation is the way to go.

If the answer to C is yes, go to D.

D. Several Critical Questions Must be Addressed

Will the system be aesthetically acceptable?
Can appropriate secondary containment for environmental protection be incorporated?
Can adequate spill control be provided to prevent fire or explosions?
Will the tank be protected from vandalism and pranks?
Can the tank system incorporate central design elements to store and transfer product?

Tank siting is critical and seldom occurs as a cookie cutter process. For instance, you won't find several ASTs dotting the landscape on the corporate campuses of Fortune 500 headquarters. Buried is better for their backup generator tanks. Similarly, the National Park Service has little interest in installing ASTs in the wild blue yonder without safeguards from vandalism, fires, theft, or other damage. If the answer to all five questions above is yes, then go to E.

If the answer to any of the five questions above is no, then redesign the system.

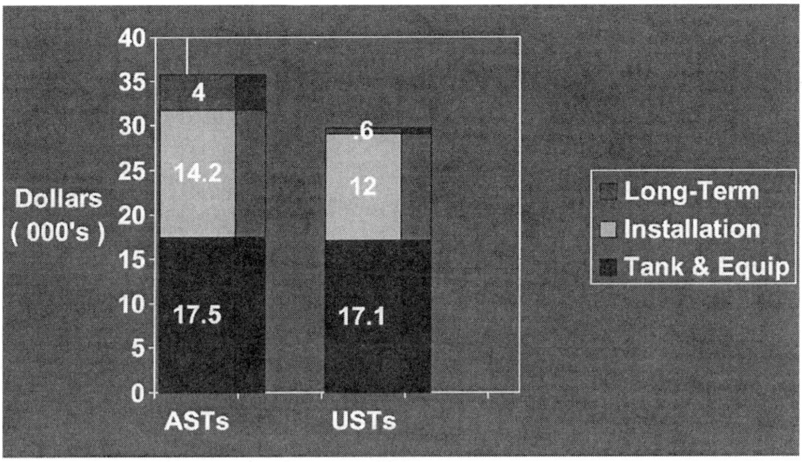

$35,700 $29,700

Figure 1 A comparison of potential AST and UST costs.

E. Is the Tank System Proposal Economically Sound? See the AST Versus UST Cost Comparison (Fig. 1)

If the answer to the above is no, then consider a UST solution or other methods (e.g., contracting for fuel services at off-site location).

If the answer to the above is yes, then develop plans, specs, and necessary contracts for installation of an AST system.

F. One Last Caution Before Beginning the AST Project

The end user also needs to be aware of "hidden" AST costs such as:

1. *Real estate*: More property is required for ASTs than USTs. An aboveground system leaves a substantial footprint on the property. A UST system operates with fewer setback restrictions than aboveground storage tanks and piping.

2. *Security*: Fencing or bollards will be needed to prevent damage from major or minor vehicle collisions. Security is addressed in both National Fire Protection Association (NFPA) codes and the U.S. Environmental Protection Agency's Spill Prevention Control and Countermeasures (SPCC) regulation. Again, this is not a one-size-fits-all question. For example, in Alaska, a fence can cause considerable problems in snow removal operations that are necessary to ensure access to an AST.

3. *Man-hours*: It takes a considerable effort to obtain permits and project approvals from state and local officials. Similarly, provisions must be made for ASTs that will require federal SPCC plans. In addition, employees who fully comprehend USTs will likely need training to better understand the unique operational requirements of AST systems.

4. *Equipment*: It's not just a tank. You may need extra piping between tank and dispenser due to fire code and building code separation requirements. Anti-siphon devices, thermal-expansion relief valves, check and block valves, and emergency vents for primary and secondary containment are also part of properly engineered AST systems.

5. *Maintenance*: A tank that's out in the open requires periodic inspection, painting, and an occasional patch job. With open dikes, the operator has to properly drain rainwater.

6. *Operational*: You may have to account for vapor losses, especially if the tank is subject to significant shifts in ambient temperature. Also, in some cases, an AST system located on a small lot—or limited to smaller capacities mandated by local zoning ordinances—will have to be filled more frequently because the overall capacity will often be less than a UST system could hold.

II. USTs

For those who have settled on the UST approach, the next question is whether a single-wall or double-wall UST system best fits the end-users' needs. Obviously, when regulations call for secondary containment, the single-wall option no longer exists. For example, certain hazardous substances—defined in Section 101(14) of the Comprehensive Environmental Response Compensation and Liability Act (CERCLA)—require secondary containment. The same is true for many non-petroleum-based chemicals as defined in 40 CFR Part 280.

When local regulations allow for flexibility, the final UST choice may depend upon the risk tolerance of tank owners. Secondary containment can provide greater assurance against environmental contamination and, in some cases, reduce insurance costs. Single-wall technology, however, has been proven reliable since the 1960s—as long as nationally recognized corrosion-protection systems are in place. Examples of recognized corrosion protection for single-wall tanks are composite construction, cathodic protection systems, and fiberglass-reinforced plastic.

Steel tanks with external cathodic protection systems must be periodically monitored. The predominant technology for cathodic protection of steel USTs is sti-P_3®.

Cathodic protection monitoring can be accomplished with a variety of methods that measure whether the anodes attached to the tank are protecting against corrosion of any metal exposed to soil. The U.S. Environmental Protection Agency requires that this monitoring—which typically takes less than 10 min to do on galvanic anode systems—be performed within six months of tank installation and at least once every three years thereafter on qualifying single-wall tanks. See Chapter 15 for more information on cathodic protection monitoring.

As part of their review of regulatory requirements, UST owners increasingly are considering whether compartmentalized USTs can provide the storage needed by their facilities. In many cases, compartmentalized designs—whether double wall or single wall—can provide dramatic savings to the tank owner. If a tank system needs to hold 20,000 gal with three different products, it's considerably less expensive to build and install one tank divided into three compartments than three USTs with the same capacity. This reduces costs by providing only one secondary containment and one interstitial monitoring device, while substantially decreasing excavation costs. In some states, tank owners can save on insurance by providing financial responsibility for one tank rather than three. However, savings are also possible by using compartmentalized secondary contained ASTs.

UST system decisions also include other equipment—piping, leak detection, overfill prevention, etc.—that will be used to provide safe handling and storage of product. See Chapter 16 for more information on UST equipment.

III. ASTs

Once the AST approach has been confirmed, the appropriate first question is, What is the application? An AST that stores liquids for manufacturing processes will often require a substantially different design than an aboveground tank that stores fuel for a school bus fleet. Bulk storage applications are considerably different than fueling because separation distances are reduced, tank bottom connections are acceptable, vertical tanks are more common, and liquids are more apt to be some sort of uncommon hazardous liquid.

After the application is properly defined, another series of questions must be addressed.

A. How Much Space Do You Have?

Particularly for fueling applications, fire codes generally dictate minimum separation distances from buildings, public ways, dispensers, other ASTs, and such. Even the type of AST allowed may be dictated by separation distance requirements.

For example, a private fleet fueling application may have plenty of room to place the AST far from roads and public access. Despite having acres of property, the business's management may decide the best location for the tank would be very close to a building. If that is the case, a protected AST would most likely be the tank of choice. Of course, this is subject to local fire code and building code provisions.

Another example: small tanks storing fuel used for remote sites, construction earth-moving activity, or a farm will face far less stringent code requirements. These tanks, if under 1100 gal, are subject to provisions in either NFPA 395 or NFPA 30A. The NFPA 395 standard:

Does not require listed tanks
Allows thinner-gauge metal
Must be separated from buildings by 40 ft
Permits dispensers to be mounted atop the tank
Incorporates emergency vents in the normal venting

However, if the remote site is subject to a local AHJ, which has established NFPA 30A as the applicable code, then the AST will confront:

More substantial setback requirements
A UL 142 tank with separate emergency vent required
The dispenser located at least 50 ft from the tank

(Any distances cited in this example would be subject to change with the use of a fire resistant tank.)

B. What if a "Fire-Rated" AST is Needed?

A fire-rated aboveground storage tank has been tested by a third-party independent lab for its ability to remain below a given maximum temperature after exposure to a 2000°F fire for a minimum of two hours. Third-party laboratories such as Underwriters Laboratories (UL) and Southwest Research Institute (SwRI) have published standards for fire-rated tanks. See Chapters 17, 18, and 19 for more information on AST tank standards.

C. What if a "Fire-Rated" AST is Not Required?

Not every storage situation will demand a tank that provides premium protection. Non-fire-rated ASTs, properly designed and installed, have been safely used around the world. Emergency vents, properly sized and operational, play a major role in assuring safe AST operation—regardless of whether the AST is fire rated.

D. What Other AST Types are There?

UL 142 is the predominant shop-fabricated aboveground storage tank standard and it includes several designs that can be adapted to provide secondary containment:

Double-wall steel AST
Vertical, horizontal, or rectangular ASTs
"Tank-in-a-box" steel AST systems—open diked or with a number of rain shield design options
Insulated double-wall steel AST
Steel AST encased in concrete
Steel AST system with piping and vents assembled at the factory

The last item is known as a UL 2244 system, which is based upon Underwriters Laboratories' Standard for Aboveground Flammable Liquids Tank Systems. The components included with the tank are vents, spill and overfill equipment, leak detection/monitoring devices, and dispensing equipment.

E. What About Tank Appurtenances?

Codes and recommended practices dwell extensively on tank system components, which play a vital role in the safe and effective operation of an aboveground storage tank. See Chapter 24 for more information on AST system equipment.

In addition to practices outlined in model fire codes, an excellent recommended practice is RP200, as published by the Petroleum Equipment Institute (PEI), which covers AST installation procedures for motor vehicle fueling applications. See Chapter 23 for more information on AST installation.

IV. THE FINISHED PRODUCT REQUIRES ATTENTION

Whether the end user chooses a system that is based aboveground or underground, the system must be checked periodically by operations or maintenance personnel.

Proper training is essential to ensure that an AST or UST is operating as designed. For example, someone has to ensure that clogged filters are cleaned or replaced, that leak detection systems are monitoring as expected, that inventory records are reconciled, and that employees and the public remain as protected as possible from the effects of a storage system release.

8
UST Design: The Wall Thickness Question

Lorri Grainawi
Steel Tank Institute, Lake Zurich, Illinois

I. INTRODUCTION

At what measured level is the steel, on a steel underground storage tank, thick enough to provide safe, long-term, reliable storage? That question was at the heart of an industry debate—and several studies—during the 1990s as Steel Tank Institute (STI) members and Underwriters Laboratories (UL) officials reexamined age-old beliefs about fabricating tanks.

Until the release of the UL 1746 (External Corrosion Protection Systems for Steel Underground Storage Tanks) standard, all steel tanks built to UL 58 required a definitive wall thickness. UL 58 (Standard for Steel Underground Tanks for Flammable and Combustible Liquids)—published in 1926—said, for example, that any steel underground storage tank from 4000 to 12,000 gal in capacity had to be built with a minimum wall thickness of one-quarter inch.

However, UL 1746 changed that wall-thickness precept upon its publication in 1989. It allowed a tank to be buried with 100 mils of fiberglass-reinforced plastic (FRP) and a steel wall thickness reduced one gauge (e.g., from ¼ in. to 3⁄16 in.) if the tank passed a special performance test. The test required that the tank be buried and subjected to a hydrostatic load equivalent to a groundwater level at grade. Also, a vacuum had to be applied without breach.

Concerns about the wall thickness issue arose shortly after the introduction of UL 1746. Industry experts began to question the risk factors and performance associated with thin-gauge tanks permitted under the new standard that governed the manufacture of composite tanks, and other emerging corrosion protection technologies.

A key departure from UL 58 was the leeway granted underground storage tank (UST) manufacturers on the thickness of steel used for the tank shell. The wall thicknesses specified in UL 58 were recognized for providing superior ductility and a structural integrity measurably better than fiberglass-reinforced plastic (FRP) underground tanks. For example, in examining the modulus of elasticity for a UL 58 tank and a comparable FRP tank, the steel tank was 30 times stiffer than the plastic version [1]. So, UL's break with tradition in allowing through UL 1746 the use of thinner-gage steel raised eyebrows within the steel tank industry.

Within the first three years of the new standard's publication, two UST tank failures were reported in which steel tanks had buckled. Both were composite tanks—employing 100 mils of fiberglass-reinforced plastic (FRP) cladding—that had been built to the reduced wall thickness allowances of UL 1746.

Until thc 1980s, the steel thickness in a UL 58 tank was recognized as a prime source of the vessel's structural integrity—and as a corrosion allowance that lengthened a tank's overall service life. The thinking, in the era before corrosion protection mandates were federal law, went like this: Thicker steel will take a longer time to perforate from corrosion than thin-gage material.

However, the thinking behind UL 1746 changed the equation. Supporters of the new standard said thin-gage steel was adequate because the superior FRP coating eliminated concerns about corrosion. In addition, the FRP would provide part of the stiffness required for long-term performance. It was also asserted that performance tests required by UL prior to granting a listing to a manufacturer would prove that UL 1746 tanks were adequately designed.

For tanks with a capacity of 10,000 gal and a diameter of 8 ft, UL 58 required a minimum nominal shell thickness of ¼-in.—or a minimum of .24-in. [2]. UL 1746 allowed two distinct minimum shell thicknesses. The standard thickness matched the UL 58 requirement. But if certain UL 1746 performance tests were passed, a reduced wall thickness was acceptable. The UL 1746 listing enabled fabrication of a 10,000-gal underground storage tank with an 8 ft diameter and a nominal steel shell thickness of 3⁄16 in.—with a minimum of .167 in. [3].

During 1990 and 1991, Steel Tank Institute members listened to reports on the two tank-buckling incidents during annual meetings of the association, which led STI to support independent research on the wall thickness question.

In collaboration with the American Iron and Steel Institute (AISI), STI funded studies by Battelle Laboratories, Dr. Reynold Watkins of Utah State University, and Dr. R. Allan Reese of Ace Tank and Equipment Co. in Seattle. Each study took a distinctive approach to the thin-gauge steel issue:

Battelle in 1991 performed a theoretical study on the additional strength provided to steel tanks of all sizes by use of 100 mils of FRP coating (1).

Watkins in 1992 studied the structural integrity of steel USTs when buried in poor, unstable soil conditions using small-scale models.

Reese in 1993 experimented to find out how various buckling pressures would affect reduced wall thickness steel USTs by studying 13 steel tanks (all of which were 4000-gal capacity).

The Reese findings were documented in a 1994 paper presented to the American Society of Mechanical Engineers' Pressure Vessel & Piping Conference in Minneapolis.

The various studies led to several key conclusions about reduced wall thickness tanks:

By reducing the tank shell thickness from ¼ in. to 3⁄16 in., the pressure at which buckling occurs drops to less than half of what the thicker shell could withstand under the test's worst-case conditions [4].

Buckling pressure can be predicted by measuring tank stiffness. There is a linear relationship between measured stiffness and measured buckling pressure (4).

Tanks fabricated with a minimum shell thickness allowed by UL 1746 (version 2), a reduced wall thickness, would be expected to fail well below 3 psi if not supported by backfill. Since pressure at a tank bottom can surpass 6 psi, backfill support is necessary to prevent thin-gage steel tanks from collapsing. Backfill is not required to prevent tanks built to the UL 58 standard from buckling, although the reserve stiffness is small for some sizes [4].

Fiberglass-reinforced plastic, as it is used to coat a UL 1746 composite steel UST, does not contribute significantly to the structural strength of the tank [1].

II. BATTELLE STUDY

In 1991, STI commissioned Battelle to investigate buckling of buried composite USTs. In this study, Battelle performed theoretical calculations, including finite element analysis. Two of the study's key findings were that the 100 mils of FRP cladding does not contribute significantly to the tank's performance, and the resistance to buckling of UL 1746 tanks featuring reduced wall thickness is less than that of UL 58 steel tanks of comparable capacity [1].

III. UTAH STATE STUDY

Because it's recognized that backfill plays an important role in preventing tank failure, STI commissioned the Watkins' study of the buckling characteristics of model tanks buried in soil. In a series of experiments, Watkins buried small-scale tanks in backfills of silty sand.

Watkins' principal conclusion was that the buckling pressure of UL 1746 tanks, varied as the fourth power of (t/D), in which t is the shell thickness and D is the tank diameter. In other words, as the stiffness of the tank decreased as measured by the ratio of steel thickness to tank diameter, the buckling pressure changed accordingly—but to the fourth power.

The result indicated that—ignoring the contribution of FRP cladding—the buckling-failure pressure for reduced wall thickness tanks would be reduced to less than one-fourth of the pressure for a comparable UL 58 tank. This was based on a 10,000-gal tank [6].

While the data from Watkins were enlightening, they were of limited value because of the complexity of the experiment, which featured variations in both the tank construction and the soil. This led STI to fund an experimental study on full-size tanks built under the same methods employed by STI members.

IV. ACE TANK STUDY

Buckling pressures were measured by collapsing 13 steel USTs through the use of water to generate external pressure. The tanks were built using minimum shell thickness specified by UL 58 and the reduced wall thickness permitted in UL 1746. The effects on buckling pressure of shell thickness and shell-to-shell weld geometry were measured, as was the effect of internal stiffeners. Buckling pressures were compared with calculated values using the Roark formula for stress and strain, which is available in technical literature.

A predeformation force was applied to the bottom of test tanks to induce uniform failure modes. A tank "stiffness" was computed from the measurements and correlated with buckling pressure. Measurements also were made on bare-steel tanks, single-wall tanks with an FRP coating, and on double-wall steel tanks [4].

V. CONCLUSIONS AND RECOMMENDATIONS

The study concluded that one of the Roark formulas for stress and strain could be used to calculate minimum requirements for steel UST wall thickness at various depths. Similarly, the study showed that engineering design calculations were superior to a performance test in predicting a tank's resistance to buckling. The study led Steel Tank Institute to the conclusion that reduced wall thickness tanks, as allowed in UL 1746, did not provide an adequate margin of safety against buckling. STI recommended that the construction of such tanks be discontinued.

"These tanks are so weak that if it were not for the backfill, it was STI's opinion that many would certainly collapse in service when subject to high water tables," according to an article in the STI newsletter, *Tank Talk*:

> The support to a tank provided by backfill depends widely on the type of backfill, how it is compacted, and on its moisture content. STI did not believe it was good engineering practice to rely on backfill to this extent. Certain types of backfill can almost turn to mud if fully saturated with water. In such a condition, backfill provides very little support to the tank [5].

The STI studies also concluded that some of the largest sizes of UL 58 tanks needed thicker-gauge steel to provide adequate safety.

STI brought its findings to UL with a recommendation to schedule an Industry Advisory Committee meeting, specifically to address shell thickness specifications for UL 58 and UL 1746. STI also presented its findings to the National Fire Protection Association committee that oversees NFPA 30 (Flammable and Combustible Liquids Code). The committee determines the adequacy of standards to be referenced within the fire code.

Driving the investigation were concerns that:

1. Reduced wall thickness tanks had an unacceptably high risk of buckling in high water tables, particularly when such tanks relied on the quality of installation to assure tank integrity.
2. STI's member manufacturers did not want to depend on the installer as the sole safeguard against potential tank failures on vessels with reduced wall thickness.
3. Manufacturers felt it was important to provide an adequate safety margin to preserve a tank's structural integrity.
4. The continuing use of thin-gauge steel in USTs presented a vast potential liability exposure for many underground storage tank stakeholders [5].

UL's response was to schedule an IAC meeting during 1995 to discuss in greater detail the findings and implications. Additional tests on 12,000-gal steel USTs were conducted in conjunction with the fact-finding mission. The Roark formula was proven to be applicable to larger tanks as well.

In March 1996, Underwriters Laboratories proposed a revision to UL 1746 that reflected some of STI's concerns [5]. The revision called specifically for the addition of the Roark equation for steel thickness. In addition, UL changed the external pressure test and eliminated several previously used evaluations such as the earth load and water load tests. UL also proposed removing the thickness of steel table from UL 1746.

UL's proposed changes to UL 1746 became effective on September 15, 1997.

REFERENCES

1. D Stephens, R Francini. Evaluation of Buckling of Buried Steel/Fiberglass Composite Storage Tanks. Battelle Laboratories, Columbus, OH 1991.
2. Underwriters Laboratories. UL 58, Standard for Safety, Steel Underground Tanks for Flammable and Combustible Liquids, 8th ed., April 15, 1986.

3. Underwriters Laboratories. UL 1746, Standard for Safety, External Corrosion Protection Systems for Steel Underground Tanks, First Edition, November 22, 1989.
4. RA Reese. Experimental investigation of buckling in full-size steel underground storage tanks. Proceedings of American Society of Mechanical Engineers, Pressure Vessels & Piping Conference, Minneapolis, June 20, 1994.
5. Steel Tank Institute. History behind proposed change of UL 58/1746 wall thickness. Tank Talk, May/June 1996, pp 1–4.
6. R Watkins. Structural Performance of Buried Steel Tanks. Utah State University, Logan UT 1992.

9

Development of UL Standards for Underground Steel Tank Safety

Shari Hunter
Underwriters Laboratories Inc., Santa Clara, California

John J. Hawley
Underwriters Laboratories Inc., Northbrook, Illinois

I. INTRODUCTION

Underwriters Laboratories Inc. (UL) evaluates more than 80,000 products in more than 17,000 categories annually, including many items for the petroleum equipment industry. UL conducts investigations on products to determine compliance with nationally and/or internationally recognized safety standards, including standards published by UL, National Fire Protection Association (NFPA), American National Standards Institute (ANSI), American Society for Testing Materials (ASTM), and the International Electrotechnical Commission (IEC).

UL Standards for Safety for steel underground storage tanks (USTs) that store flammable and combustible liquids include (1) UL 58 (Steel Underground Tanks for Flammable and Combustible Liquids), and (2) UL 1746 (External Corrosion Protection Systems for Steel Underground Storage Tanks). UL also has published UL 1316, which governs the manufacture of USTs fabricated from fiberglass-reinforced plastic (FRP).

UL standards are developed based on the needs of state and federal regulatory authorities. During the development of safety standards, UL solicits advisory information from regulators, manufacturers, trade organizations, and others interested in the specific standard.

II. HISTORY OF UL'S UNDERGROUND TANK STANDARDS

The first Industry Advisory Conference (IAC) meeting for development of UL 58 for steel underground tanks was convened at UL's Chicago office on March 31, 1919. The first edition of UL 58 was published in 1925.

With the United States assisting besieged European nations prior to World War II, an Emergency Alternate Specification (EAS) was issued on October 1, 1941, based on the limited availability of galvanized steel. At this time, all underground steel tanks fabricated of material less than 7 USS gage were required to be galvanized steel. This requirement applied to all tanks up to a maximum capacity of 1100 gal. The EAS allowed use of No. 10 gage black steel as an alternative to galvanized steel. A special label was developed to identify such tanks and specified the letters "EAS," No. 10 USSg, and the date of manufacture. On April 1, 1949, UL 58 was revised to allow the use of either galvanized steel or heavier gauge uncoated steel for tanks up to 1100 gal.

The May 1957 edition of NFPA 30 (Flammable Liquids Code) called for a minimum steel thickness on underground storage tanks of:

14 gage for tanks up to 285 gal capacity
12 gage for tanks up to 560 gal capacity
10 gage for tanks up to 1100 gal capacity
7 gage for tanks up to 4000 gal capacity
One-quarter inch for tanks up to 12,000 gal capacity

Tanks carrying the UL label were considered as meeting the steel thickness requirement [1].

Based upon data referenced in UL's July 20, 1959, bulletin, the next draft of UL 58 specified the following steel thicknesses—whether the metal was galvanized or uncoated:

14 gage for tanks up to 285 gal capacity
12 gage for tanks up to 560 gal capacity
10 gage for tanks up to 1100 gal capacity
7 gage for tanks up to 4000 gal capacity
One-quarter inch for tanks up to 12,000 gal capacity
$\frac{5}{16}$ in. for tanks up to 20,000 gal capacity
$\frac{3}{8}$ in. for tanks up to 50,000 gal capacity

Until 1959, underground steel tanks were required to have a coat of tar, durable asphaltic-base paint, or red lead and oil paint. In 1961, UL 58 was revised to require tanks to have only a coat of paint to retard corrosion during storage. Since the level of corrosion protection required for tanks was at the discretion of the regulatory authority, it was often necessary to remove the factory paint to upgrade the corrosion protection. For the asphaltic type of coatings, removal often proved to be difficult, necessitating the use of steam, sandblasting, or other techniques. Since it was seldom known where the tank would be installed or the level of corrosion protection required, manufacturers proposed that the requirement for painting be waived. UL accepted the manufacturers' proposal.

However, it was not until 1968 that the UL 58 mandate to paint the tank was deleted since NFPA specified the corrosion protection required. NFPA 30 required cathodic protection or corrosion-resistant materials such as special alloys, fiberglass-reinforced plastic (FRP) or FRP coatings, if deemed necessary by a qualified engineer. Selection of the type of protection to be employed was to be based on the corrosion history of the area and the judgment of a qualified engineer, subject to approval of the fire inspector or other authority having jurisdiction. The factory coating was intended only as a protection against atmospheric corrosion during storage and transit. This coating often consisted of a single coat of paint or primer. The purpose of this factory coating was not always understood since the paint could be misconstrued in the field as providing adequate corrosion protection.

In fairness to fire inspectors of that era, a vast body of knowledge has developed during the last three decades on steel tank corrosion—its environmental impacts, and the best ways to counteract it. As innovative ideas surfaced, tank manufacturers and corrosion experts advanced an array of reliable solutions that include durable, dielectric materials, cathodic protection, and many forms of secondary containment for underground storage. Further, the fire inspector's primary concern regarding fuel storage was the fire hazard, which was essentially mitigated by burying the tank. It wasn't until many years later that concern for environmental issues came into play.

On May 11, 1983, UL convened an ad hoc committee meeting to discuss topics related to underground tanks, including corrosion protection for steel tanks. On December 8, 1983, UL convened an IAC meeting at which the issue of corrosion protection was discussed. Based on information provided, UL proposed to develop a Standard for Corrosion Protected Tanks.

The proposal for a new standard was also prompted by a change in NFPA 30. Prior to 1983, NFPA 30 required corrosion protection only if deemed necessary by a qualified engineer. In 1983, that requirement was revised to specify corrosion protection unless waived by a qualified engineer. During the December 8, 1983 IAC meeting there was also considerable discussion on the subject of secondary containment tanks. Using input received, UL developed requirements for secondary containment tanks. On March 30, 1984, UL issued a bulletin indicating that such tanks may be submitted for investigation and listing. UL initially defined a double-wall tank as two tanks in one (e.g., a complete tank enclosed within a secondary tank) and provided with a means for leak detection in the annulus—the space between the inner and outer tank. UL subsequently considered constructions referred to as "wrapped tanks." The wrapped tank is essentially a UL 58 tank with a steel intimate-contact outer jacket encircling a portion of a tank circumference. The area between the inner tank and outer jacket also can be monitored.

With the increasing emphasis on protection of groundwater, a number of regulatory authorities suggested that UL evaluate field lining of leaking steel underground tanks. Such lining was considered to be a viable option, especially in situations where removal and replacement were impractical. After considerable review and discussions with lining applicators, material suppliers and regulatory authorities, UL determined that such a program was feasible and the requirements were included in the May 1987 Subject 1856 Bulletin (Outline of Investigation Used to Evaluate Tank Lining Systems).

III. UL 58

UL 58 covers single- and double-wall steel underground tanks. In 1925, the requirements for steel underground tanks were transferred from the regulations of the National Board of Fire Underwriters (NBFU) for the Installation of Containers for Hazardous Liquids and published in the first edition of UL 58. In 1963, the installation regulations in the NBFU were transferred to NFPA 30. UL 58 addressed safety issues relating to leakage, buckling from external pressure and pressure venting. The tables in the original UL 58 identified the minimum thickness of the steel based on the tank diameter and maximum capacity of the tank. Performance testing was not required to evaluate the design of the tank.

The edition of UL 58, published on December 13, 1996, allows manufacturers to determine the minimum steel shell thickness based on one of the Roark equations on stress and strain. If the tank is designed in accordance with the Roark equation, performance testing is not required to evaluate the design of the tank.

$$t_{\mathrm{s\,min}} = [(P_1 L r^{3/2}(1 - u^2)^{3/4})/(.807E_s)]^{.4}$$

where P_1 = calculated external pressure (in psi) at the bottom of a tank submerged in water to a depth of 5 ft or the maximum burial depth specified by the manufacturer; E_s = modulus of elasticity for the steel; t_s = thickness of steel tank shell (inches) with a minimum thickness of 0.123 in. for the primary tank and 0.093 in. for the secondary tank; L = length of the tank (inches); r = radius of the tank (inches); and u = Poisson's Ratio.

UL 58 permits the steel thickness of the tank to be reduced up to 25 percent—across the full range of capacities—provided the design withstands an external pressure test. During the one-hour external pressure test, the tank is submerged in water to a depth equal to the manufacturer's specified burial depth or 5 ft, whichever is greater. The tank is not permitted to leak, collapse, implode or buckle (defined as deflection of 5 percent of the tank diameter) during this test.

UL 58 also includes requirements for lifting lugs to be provided with the tank. The lifting lugs are required to support, for one second, a load equal to two times the empty weight of the tank without permanent deformation and the tank shall not leak when subjected to a 5 psi pressure test.

UL 58 specifies certain types of welds, 144-in. maximum diameter, a maximum length-to-diameter ratio of 8 to 1, and the minimum vent size. During production, each tank must be tested for leakage before shipping or coating.

IV. UL 1746

This standard was first published in November 1989. It provides requirements for corrosion protection of steel underground tanks. The base tank is constructed in accordance with UL 58 and corrosion protection is then added. The types of protection systems include:

Cathodically protected tanks
Composite tanks
Jacketed tanks

Table 1 outlines the tank tests UL conducts before a manufacturer can obtain a UL 1746 listing.

A. Part I—Preengineered Systems

Part I of UL 1746 covers preengineered cathodic protection systems. The tanks are provided with galvanic anode(s). The material specifications are in UL 1746. It also specifies requirements for the backfill around the anode, insulating bushings for electrically isolating the tank from the piping, and wire connectors for cored anodes. The cathodic protection system is constructed to provide a minimum system design based on a 4000 ohm-cm soil resistivity.

B. Part II—Composite Tanks

Part II of UL 1746 covers composite tanks. These are steel tanks complying with UL 58 that are coated with a nonmetallic material at least 0.100 in. thick. The steel thickness of the tank may be reduced if the earth load test, water load test, and external pressure test are conducted on a representative sample tank.

C. Part III—Jacketed Tanks

A jacketed tank is subjected to an interstitial communication test during which a sample tank is installed underground and filled to capacity with water. The test be-

Table 1 UL 1746 Tests

Tests required	Part I	Part II	Part III	Part IV
Anode	×			
Isolating devices	×			
Lift lugs	×	×	×	×
Strength of pipe Ættings	×	×	×	×
Tank impact		×	×	×
Tank holiday test		×	×	×
Interstitial communication			×	
Annulus proof pressure			×	
Coating tests				
Environmental Øuids	×	×	×	×
Oven aging	×	×	×	×
Light and water exposure	×	×	×	×
Cold impact	×	×	×	×
Abrasion resistance	×			
Cathodic disbondment	×			
Corrosion evaluation		×	×	×
Permeation		×	×	×
Fuels			×	
Flexibility	×			

Source: UL 1746.

gins on the end of the tank as far as possible from the interstitial monitoring point. Water is introduced to an atmospheric annular area to simulate a leak through the outer wall of a vacuum-monitored interstitial space. There shall be measurable communication to the monitoring point within 24 hr.

If a jacketed tank is constructed with reduced steel thickness, the water load test, the earth load test, and the external pressure test are conducted on a tank assembly.

D. Part IV—Urethane-Coated Tanks

A fourth part to UL 1746, which will apply to tanks with thick-film urethane coatings, was under development at the time of publication.

V. UL STANDARDS DEVELOPMENT PROCESS

UL seeks approval of the Standards for Safety by the American National Standards Institute (ANSI), whose approval indicates national recognition for the scope of products covered by the standard. ANSI approval assures that due process was utilized in development of the standard, and consensus was achieved by involving all interested organizations.

The ANSI approval process includes three methods to achieve consensus. The standards development organization (SDO) is to be accredited by ANSI to use one or more of the following three methods:

1. *Organization Method.* This requires the SDO to write its own procedures, which must meet the due process requirements on the Procedures for Development and Coordination of American National Standards ("ANSI Procedures"). This method permits flexibility since the SDO can develop standards within the structure and procedures of its own organization.

2. *Committee Method.* This requires formation of a committee (i.e., a consensus body) to develop a draft standard on which consensus must be reached before it's submitted to ANSI for approval. The committee is required by the ANSI Procedures to consist of directly and materially affected interests; it serves as a forum for these different interests to be represented. The committee usually adopts the model procedures in Annex A of the ANSI Procedures. As an alternative, it may develop its own procedures consistent with the ANSI Procedures. The committee membership is to be balanced, and cannot be dominated by any single interest category. Any subgroup (i.e., a working group actually writing the draft standard) is not required to be balanced.

3. *Canvass Method.* This requires a ballot to determine consensus. When using the canvass method, the SDO uses its established procedures to develop the standard. In this case, the organization is comparable to the subgroup formed under the Committee Method. The SDO attempts to resolve documented objections, and report the results and resolutions to ANSI. The process in Annex B of the Procedures for the Development and Coordination of American National Standards is to be used by an organization employing the canvass method. The UL/ANSI steel tank standards were approved under the canvass method.

VI. CONCLUSION

During the 75 years that UL has been developing steel tank standards, the past two decades have been the most challenging. As technology continues to change, UL standards will evolve to address the latest innovations.

REFERENCE

1. National Fire Protection Association. Flammable Liquids Code, NFPA 30. 1957.

10

Development of ULC Standards for Underground Storage and Handling of Flammable and Combustible Liquids

Gordana Nikolic
Underwriters' Laboratories of Canada, Scarborough, Ontario, Canada

I. INTRODUCTION

Underwriters' Laboratories of Canada (ULC) is a not-for-profit organization with headquarters in Scarborough, Ontario, Canada. It is a Canadian safety, certification, testing, quality registration, and standards development organization.

Established in 1920 by Underwriters Laboratories Inc. (UL) as an independent third-party product certification organization, ULC has built its mark into one of Canada's most respected and widely accepted symbols of product safety for construction materials, fire protection and suppression products, burglary protection, and signaling systems.

On August 15, 1920, ULC was granted a charter under a letter patent issued by the Secretary of State of Canada. Since then, ULC has operated as a link between Canadian regulatory authorities, insurance inspection agencies, manufacturers of products and suppliers of services. The Standards Council of Canada has accredited ULC as a testing, certification, quality registration, and standard development organization under the National Standards System of Canada.

On September 21, 1995, UL and ULC entered into a formal corporate affiliation that provides a comprehensive vehicle for manufacturers to enjoy "one-stop shopping" at either organization to gain any of the UL and ULC listing marks.

II. CANADIAN STANDARDS DEVELOPMENT

ULC has been certifying and developing standards for flammable and combustible liquid storage products for more than 50 years. ULC certification of tanks has been a mainstay in the business and continues to be a major part of ULC's operation.

Initially, there were only four types of listed tanks: underground, aboveground horizontal, aboveground vertical, and indoor tanks for oil burners. ULC standards for those tanks were straightforward and prescriptive. The standards focused on the new product (i.e., the materials, fittings, methods of assembly, and factory leak testing). At that time, the standards' primary focus was to ensure that tanks were leak-tight when they left the factory.

During the past three decades, the trend in Canadian standards development shifted as stakeholders began demanding that committees, rather than individual organizations, write standards. These committees were required to have balanced representation from end users, manufacturers, inspection authorities, research organizations, testing labs, consultants, etc. In response, Underwriters' Laboratories of Canada formed two committees: one to develop steel tank standards and the other to develop requirements for fiberglass-reinforced plastic underground tanks. The inaugural meeting of ULC Committee on Stationary Steel Storage Containers for Flammable Liquids, S600A met on June 18, 1974, followed by ULC Committee on Stationary Nonmetallic Storage Containers for Flammable Liquids, on December 15, 1975. Both committees remain very active to this day.

The recognition of the potential hazards associated with tank leaks has resulted in more stringent requirements for underground storage tanks to reduce the likelihood of such incidents. During the past five years, heightened environmental awareness has influenced development of many new standards that affect the storage of flammable liquids. Consequently, there has been a trend to replace traditional underground storage systems with aboveground storage, where the tank is readily available for inspection and any leakage can be detected visually before significant quantities of product could be released to the environment. Chapter 19 will address ULC's efforts at providing standards for aboveground storage systems.

Responding to increasing concerns about protection of the environment and public safety through the use of underground storage tanks, the Steel Tank Committee generated a number of standards for underground tanks and accessories. The published standards are:

CAN/ULC-S603 (Steel Underground Tanks for Flammable and Combustible Liquids)

CAN/ULC-S603.1 (Galvanic Corrosion Protection Systems for Steel Underground Tanks for Flammable and Combustible Liquids)

ULC-S616 (Testing of Liquid Protective Coating Materials as Required by CAN/ULC-S603.1 for use in connection with the Corrosion Protection of the Underground Tanks)

CAN/ULC-S618 (Magnesium and Zinc Anode Assemblies and Zinc Reference Electrodes)
CAN4-S631 (Isolating Bushings for Steel Underground Tanks Protected with Coatings and Galvanic Systems)
ULC-S652 (Tank Assemblies for Collection of Used Oil)

The committee on nonmetallic tanks has produced one standard, ULC-S615 (Reinforced Plastic Underground Tanks for Flammable and Combustible Liquids).

To address technological change, ULC's committees are continuously reviewing proposed modifications to existing standards. The revised requirements and new amendments are prepared and incorporated into new editions of ULC standards. While many benefits are realized from a consensus-based standards development system, it should be noted that consensus standards can take a long time to produce. Depending on the complexity, two years or more is a common interval from conception to publication of a formal standard. This can cause problems for both regulators and industry officials because new concepts and products are continuously under development. Consequently, ULC recognizes an ongoing need to develop certification criteria quickly so that products can get to market in a timely fashion. ULC, under its accreditation by Standards Council of Canada (SCC), can develop and publish Other Recognized Documents (ORDs) to establish certification criteria. These ORDs are circulated to relevant regulatory authorities for approval and adoption. This entire process can be completed in about three to four months.

ULC has prepared a number of ORDs in these areas and is developing new ORDs for specialized underground tanks and accessories. The following list contains published ULC ORDs and ORDs under development.

ULC/ORD-C58.10 (Jacketed Steel Underground Tanks for Flammable and Combustible Liquids)
ULC/ORD-C58.20 (Special Corrosion Protection Underground Tanks for Flammable and Combustible Liquids)

In the fields of leak prevention and detection, ULC was requested by the National Task Force of the Canadian Council of Ministers of the Environment (CCME) to develop a series of requirements reflecting environmental safety. Collectively they are as follows:

ULC/ORD-C58.9 (Secondary Containment Liners for Underground and Aboveground Flammable and Combustible Liquid Storage Tanks)
ULC/ORD-C58.12 (Leak Protection Devices [Volumetric Type] for Underground Flammable and Combustible Liquid Storage Tanks)
ULC/ORD-C58.14 (Non-Volumetric Leak Detection Devices for Underground Flammable Liquid Storage Tanks)
ULC/ORD-C58.15 (Overfill Protection Devices for Flammable Liquid Storage Tanks)

ULC/ORD-C58.19 (Spill Containment Devices for Underground Flammable and Combustible Liquid Storage Tanks)
ULC/ORD-C107.4 (Ducted Flexible Underground Piping Systems for Flammable and Combustible Liquids)
ULC/ORD-C107.7 (Glass-Fibre Reinforced Plastic Pipe and Fittings for Flammable and Combustible Liquids)
ULC/ORD-C107.12 (Line Leak Detection Devices for Flammable Liquid Piping)
ULC/ORD-C107.19 (Secondary Containment of Underground Piping for Flammable and Combustible Liquids)
ULC/ORD-C107.21 (Under Dispenser Sumps)
ULC/ORD-C107.14 (Underground Piping)
ULC/ORD-C180 (Liquid Level Gages and Indicators for Fuel Oil and Lubricating Oil Tanks)
ULC/ORD-C586 (Flexible Metallic Hose)

Because ORDs are approved and adopted by Canadian authorities, they are accepted and enforced throughout the country. Many of the ORDs and standards previously listed have also been incorporated into national and provincial codes and other regulations.

III. UNDERGROUND TANKS

It has been estimated that there are 200,000 underground tanks installed across Canada. There are national regulations that address the design, installation and monitoring of underground storage facilities.

Underground storage tanks are of nonpressure type intended for the storage of flammable or combustible liquids, such as gasoline, fuel oil, or other similar products. These products are listed under ULC's label service program. After the certification investigation is successfully completed, and a listing promulgated, periodic examinations and tests are conducted on samples selected at random from current production and stock.

Currently, ULC has three listing categories for underground tanks. The remainder of this section addresses specific tank types and their associated standards and ORDs.

During the past 10 years, concerns have escalated regarding the potential for leakage from underground tanks. The main cause of tank leakage has been the corrosion of unprotected steel tank walls due to naturally occurring electrical potential gradients between the metal and moisture-laden soil. In a small number of cases, internal tank corrosion has occurred when water accumulated inside the tank.

Public environmental awareness created a need to develop a new set of standards and regulations, which eventually led to a general upgrading of products as well as new product development. New designs of underground tanks emerged:

Double-wall tanks with the ability to monitor an interstice

Jacketed tanks (e.g., underground steel tanks with a fiberglass outer shell creating an interstice that can be monitored to detect leakage from the primary tank)

Special purpose cathodically protected tanks (e.g., steel tanks externally coated with nonmetallic materials to a minimum thickness of 2.5 mm)

In response to the new generation of underground tanks, ULC developed and published new documents to address their needs for performance and quality standards (Table 1).

In 1983, ULC published the first edition of CAN4-S615 (Standard for Reinforced Plastic Underground Tanks for Petroleum Products). In 1985, the ULC Standard Committee on Stationary Steel Storage Containers for Flammable Liquids published two standards: CAN/ULC-S603 (Standard for Steel Underground Tanks for Flammable and Combustible Liquids) and CAN/ULC-S603.1 (Standard for Galvanic Corrosion Protection Systems for Steel Underground Tanks for Flammable and Combustible Liquids). In addition, ULC has developed ULC/ORD-C58.10 (Jacketed Steel Underground Tanks for Flammable and Combustible Liquids) and ULC/ORD-C58.20 (Special Corrosion Protection Underground Tanks for Flammable and Combustible Liquids).

A. CAN/ULC-S603 (Standard for Steel Underground Tanks for Flammable and Combustible Liquids)

This standard covers storage tanks, of the nonpressure type, that are intended for the underground storage of flammable and combustible liquids such as gasoline, oil, and similar products. The requirements provide construction-based design information, such as minimum steel thickness, maximum diameter, diameter-to-length ratio, weld joints and minimum vent openings. These tanks may be single-wall, double-wall or compartmentalized designs. Double-wall tanks include a secondary containment that covers a minimum of 300° of the circumferential surface area of the primary tank and 100 percent coverage of the primary tank heads. Construction of the secondary containment is separate from, but includes attachment to, the primary tank by stitch welding along the circumferential edges of each shell plate. This provides an interstice between the primary tank and containment. The remaining area at the top of the tank is single wall where the fittings are located. These tanks may be provided with manhole access.

Table 1 Underground Tanks Terms (Underwriters' Laboratories of Canada)

Standard/ document	Type	Description	Comments
	Underground tanks (ULC Guide No. 60 O5.1)		
CAN/ULC-S603	Underground steel tank	Single, double, compartment	For Øammable and combustible liquids
CAN/ULC-S603.1	Galvanic-protected underground tank	Single, double, compartment	
CAN4-S615	Reinforced-plastic underground tank	Single, double, compartment	
	Jacketed underground tank (ULC Guide No. 60 O5.10)		
ULC/ORD-C58.10	Jacketed steel tank	Encapsulated S603 tank	For Øammable and combustible liquids
ULC/ORD-C58.20	Special corrosion protection tank		
	Tanks for used oil (ULC Guide No. 60 O5.03)		
ULC-S652	Tank for storage of used oil	Single, double, contained, encased	ModiÆed underground and aboveground tanks

All single- and double-wall tanks must be leakage tested by applying pressure to the primary tanks as specified in the CAN/ULC-S603 Standard. All double-wall tanks are required to be leak tested by drawing required vacuum on the interstice. The secondary containment tanks are factory equipped with permanent vacuum monitoring devices. Double-wall tanks shall be protected from external corrosion in accordance with standard CAN/ULC-S603.1.

B. CAN/ULC-S603.1 (Galvanic Corrosion Protection Systems for Steel Underground Tanks for Flammable and Combustible Liquids)

These requirements cover galvanic protection systems for cylindrical horizontal steel tanks of nonpressure type manufactured in accordance with the CAN/ULC-S603 standard. In general, this standard provides requirements for construction of galvanic protected underground tanks as follows:

Tank connections must be made with bushings or flanged fittings with isolating gaskets and sleeves. The bushing and sealants must comply with the requirements of CAN4-S631 (Standard for Isolating Bushings for Steel Underground Tanks Protected with Coatings and Galvanic Systems)

Where access to a manhole or pumping equipment is required through a riser, the tank is kept electrically isolated from its surrounding environment, including the attached piping.

Each tank is factory equipped with a sacrificial anode system complying with the applicable requirements of CAN/ULC-S618 (Standard for Magnesium and Zinc Anode Assemblies and Zinc Reference Electrodes)

C. Other Underground Tanks

The following standards cover other types of underground tanks and methods of preventing corrosion. These standards are performance based, and address issues such as compatibility of nonmetallic materials with contained fuels and external load testing (e.g., flood load and burial load test).

ULC-S615 (Standard for Reinforced Plastic Underground Tanks for Flammable and Combustible Liquids). These are nonmetallic tanks, predominantly of cylindrical design, and are of the nonpressure type. They may be of single- or double-wall construction. The walls of the primary tank and secondary containment are separated by a narrow gap. Provision is made for monitoring interstitial space by pressure, vacuum, or mechanical means.

ULC/ORD-C58.10 (Jacketed Steel Underground Tanks for Flammable and Combustible Liquids). Jacketed tanks consist of primary steel tanks that meet the requirements of CAN/ULC-S603 with a secondary containment made of non-metallic materials (mainly fiberglass-reinforced plastic). The secondary containment completely encapsulates the primary tank to form an interstitial space.

ULC/ORD-C58.20 (Special Corrosion Protection Underground Tanks for Flammable and Combustible Liquids). These are steel tanks meeting the requirements of CAN/ULC-S603 that are coated with the nonmetallic material at least 100 mils in thickness. These standards require the attachment of one anode to the tank to protect it from corrosion if a holiday is undetected in the dielectric coating.

IV. CONCLUSION

Heightened environmental awareness has influenced development of many new products. As a result, the storage of flammable and combustible liquids has been altered by many recent innovations. Many new products are entering the marketplace. ULC, as a Canadian certification and standards development organization, is working closely with authorities and manufacturers to develop requirements for products and systems that ensure public safety remains paramount while meeting market needs.

The future will bring ongoing revisions to existing standards and completely new standards addressing new generations of products. End users will continue to rely on the ULC and cUL marks to ensure products meet the most current requirements in the field of flammable and combustible liquid storage.

11
Federal UST Regulatory Program

Marcel Moreau
Marcel Moreau Associates, Portland, Maine

I. WHY REGULATE UNDERGROUND STORAGE TANKS AND PIPING?

The leaking underground storage system problem simmered for a long time before boiling into a national regulatory program. As early as 1941, the National Fire Protection Association (NFPA) had warned that leaking tanks were a large problem and that measures such as corrosion protection and periodic testing of tanks should be implemented (1).

The 1950s saw vast increases in the number of storage systems installed, with an average of 5000 to 7000 new systems per year being placed in operation from the early '50s to the early '70s (2). By the mid-1960s, petroleum marketers became aware that corrosion was a problem (3). The industry began to examine and use new technologies such as fiberglass tanks and steel tanks with factory-installed corrosion protection systems.

Still, the petroleum industry failed to wholeheartedly endorse the new technologies, and the problem continued to grow. By the late 1970s some petroleum marketing companies began to take steps to assess the magnitude of the problem. Within a few years studies began to reveal that the problem was indeed large—the leaking tank population was estimated to include perhaps 70,000 tanks with as many as 350,000 leaking within the next five years (4).

At the same time, a few well-publicized and very expensive contamination cleanups (and associated lawsuits) resulting from underground storage system leaks made the petroleum industry acutely aware of leaks and their potential financial consequences (4). By the mid-1980s, local and state level regulations also began to proliferate, imposing a patchwork of regulatory requirements upon storage system owners.

Congress conducted hearings on the tank problem during the fall of 1983, revealing the general absence of any environmentally based standards for USTs (5). By early 1984, federal legislation was introduced that would create a national underground storage system regulatory program. The bill sped through Congress and was signed into law on November 8, 1984, by President Ronald Reagan.

The federal legislation created a tank program within the Resource Conservation and Recovery Act that was to be implemented by the U.S. Environmental Protection Agency. Over the next four years the EPA would struggle to design a program that would eventually govern the activities of some two million storage systems owned by 700,000 different owners (6). The main components of the program are outlined in this chapter.

II. WHAT IS REGULATED

The main thrust of the federal program is to regulate underground petroleum storage tanks containing used oil, motor fuels, or hazardous liquids used for commercial purposes. Heating-oil supply tanks of any capacity connected to oil burners were exempted from the regulations, as were farm and residential motor fuel tanks of 1100 gal or less. Emergency generator tanks are regulated, except they are not required to meet leak detection requirements. For a more complete description of the tanks that are regulated and exempted, refer to 40 CFR 280.10 and 280.12.

A fundamental tenet of the federal program is that state and local regulations can be more stringent. For example, most of the northeastern states, where fuel oil is commonly used for heating buildings in winter, have imposed some requirements on heating-oil supply tanks. Some states do require leak detection on emergency-generator tanks. Check with your state implementing agency (Appendix) for exact requirements of the state in which your storage system is located.

III. WHAT IS REQUIRED

The federal rules' requirements were designed to fix the sins of the past and anticipate the sins of the future. The main elements of the rules are the following.

A. Notification

When the federal law went into effect, no one really knew exactly how many storage systems were in existence, where they were, or who owned them. To correct this deficiency, the federal law included a requirement for all regulated tank owners to notify their state agency that they owned tanks. While the federal law only provided for a one-time notification of tank ownership, all states have expanded this program into an ongoing registration and/or permitting program so that an up-to-date database of tank owners, tank locations and basic tank information can be maintained.

B. Installation

Because the bare-steel storage system had been identified as a prime source of leaks in industry (7) and U.S. EPA surveys of the 1980s, the federal law included an "interim prohibition" on the continued use of these materials for newly installed storage tanks, piping, and related equipment. This was a practical step because it recognized that a comprehensive set of regulations would take several years to formulate. During that time, there was no reason to repeat the mistakes of the past while waiting for detailed regulations to be written and implemented. The interim prohibition specified that after May 7, 1985, all newly installed storage systems must be protected against corrosion, structurally sound and compatible with the tank contents.

The interim prohibition was supplanted when the federal rules went into effect on December 22, 1988. The regulations specified the types of storage systems that were acceptable, including cathodically protected steel, steel clad with non-corrosive material, and fiberglass. The final regulations did not include specific requirements for how new systems were to be installed. However, the new program did include requirements to follow industry practices (such as the Petroleum Equipment Institute's *Recommended Practices for Installation of Underground Liquid Storage Systems* and the American Petroleum Institute's *Installation of Underground Petroleum Storage Systems*) and manufacturers' recommendations during storage system installation.

To deal with spillage associated with deliveries to storage systems, new tanks were required to be equipped with devices to contain small drips and spills around the fill pipe and prevent the storage tank from being overfilled.

C. Operation

To promote the early detection of releases, the federal regulations included requirements for leak detection of tanks and piping. Leak detection requirements were to be imposed immediately on newly installed tanks, and were to be phased in for existing storage systems on a timetable based upon a tank's age. By December 22, 1993, all in-service storage systems were required to have leak detection technology in place.

The philosophy behind the leak detection regulations was to permit as many proven technologies as possible. This approach recognized the vast variety of storage systems in existence and was intended to promote the development of as many different leak detection approaches as possible.

The list of leak detection options described in the regulations for tanks included:

Monthly inventory control used in conjunction with periodic tank tightness testing
Weekly manual tank gaging, limited to small tanks of 2000 gal capacity or less
Monthly automatic tank gauging
Monthly monitoring of secondary containment systems
Monthly monitoring of groundwater for signs of contamination
Monthly monitoring of soil vapors for signs of contamination

Leak detection options for piping included:

Periodic tightness testing
Monthly monitoring of secondary containment systems
Monthly monitoring of groundwater for signs of contamination
Monthly monitoring of soil vapors for signs of contamination
Leak-proof construction (suction systems only, also known as "safe suction" or "European suction")

Because of hazards posed by leaks from fuel delivery piping that operated under pressure, these pumping systems were also required to have automatic line leak detectors.

The rules also contained provisions for other methods of leak detection, if the technology was able to detect leaks of 0.2 gal/hr or 150 gal/mo. The only widely used leak detection method that comes under this section of the rule is statistical inventory reconciliation, a method that can be applied to both tank and piping leak detection.

D. Upgrading

Rather than require that all existing storage systems that did not meet the current requirement for corrosion protection be replaced, the regulations allowed for "upgrading" such systems to meet the regulations. Upgrading technologies included internal tank lining and the addition of cathodic protection. Spill containment around fill pipes and overfill prevention equipment were also to be retrofitted. There was a single deadline for this upgrading, 10 years after the regulations took effect, on December 22, 1998. After this milestone in the regulatory program, all in-service storage systems were to finally have corrosion protection, spill containment, and overfill prevention—as well as ongoing leak detection.

E. Closure

Prior to federal regulation, many storage systems were left in place after they were no longer used. Leakage and spillage around these abandoned systems were seldom addressed. Even when tanks were removed and replaced, cleanup of contamination was cursory at best. Federal closure regulations were intended to ensure that out-of-service tanks were properly closed, and that any contamination associated with the storage system be assessed and remediated.

Regulations require notification of the regulatory agency when a tank owner or manager intends to close a storage system. In addition, they must file a formal contamination assessment of the area around the storage system.

F. Financial Responsibility

To ensure that resources were available to deal with leaks, spills, and resulting contamination, regulations required that tank owners document their ability to pay for cleanups and damages that might be caused by operation of the storage system. Financial responsibility could be demonstrated through a number of mechanisms such as private insurance, letters of credit, or self-insurance for large companies with significant financial resources. Most states eventually set up funds that provided financial responsibility for the bulk of storage system owners in the state.

REFERENCES

1. National Fire Protection Association. Leakage from Underground Gasoline Tanks. Quincy, MA: NFPA 30B, Flammable Liquids Ordinance, 1941, pp 40, 43.
2. H Upton. Industry outlook. Proceedings of Petroleum Equipment Institute Convex '79, Boston, September 25, 1979.
3. Solving the leaking tank problem. National Petroleum News April 1966, pp 128–132.
4. The tank leak mess. National Petroleum News July 1982, pp 37–40.
5. Leaking Underground Storage Tanks: A Potential Environmental Problem. Washington: Congressional Research Service, January 11, 1984, p 16.
6. Federal Register. September 23, 1988, p 37095.
7. Tank and Piping Leak Survey. Washington: American Petroleum Institute, 1981, p 7.

APPENDIX: STATE AND U.S. TERRITORIES UNDERGROUND STORAGE REGULATORS

Alabama
Alabama Dept. of Environmental Management
Groundwater Section/Water Division
1751 Congressman W.L. Dickinson Dr.
Montgomery, AL 36130

Alaska
Alaska Dept. of Environmental Conservation

Storage Tank Program
410 Willoughby Ave.
Juneau, AK 99801-1795

Arkansas
Arkansas Dept. of Pollution Control & Ecology
Regulated Storage Tanks
P.O. Box 8913
8101 Interstate 30, Bldg. D
Little Rock, AR 72219-8913

Arizona
Arizona Dept. of Environmental Quality
UST Programs and Support Section
Inspections and Compliance Unit
3033 N. Central Ave.
Phoenix, AZ 85012

California
California State Water Resources Control Board
Division of Clean Water Programs
P.O. Box 944212
Sacramento, CA 94244-2120

Colorado
Colorado Department of Labor and Employment
Oil Inspection Section
1515 Arapahoe St.
Tower 3, Suite 610
Denver, CO 80202-2117

Connecticut
Connecticut Dept. of Environmental Protection
Waste Management Bureau
State Office Bldg.
79 Elm St.
Hartford, CT 06106

Delaware
Delaware Dept. of Natural Resources and Environmental Control
Division of Air and Waste Management
UST Branch
391 Lukens Dr.
New Castle, DE 19720-2774

District of Columbia
DC Environmental Health Administration
Pesticides, Hazardous Waste & UST Division
2100 Martin Luther King Jr. Ave., SE
Suite 203
Washington, DC 20020

Florida
Florida Dept. of Environmental Regulation
Tank Section
Twin Towers Office Bldg.
Room 403
2600 Blair Stone Rd.
Tallahassee, FL 32399-2400

Georgia
Georgia Dept. of Natural Resources
UST Management Program
4244 International Parkway
Suite 104
Atlanta, GA 30354

Hawaii
Hawaii Dept. of Health
Solid and Hazardous Waste Branch
919 Ala Moana Blvd.
Room 212
Honolulu, HI 96814

Idaho
Idaho Division of Environmental Quality
1410 North Hilton
Boise, ID 83706

Illinois
UST Contact:
Illinois Office of State Fire Marshal
Division of Petroleum and Chemical Safety
1035 Stephenson Dr.
Springfield, IL 62703
LUST Contact:
Illinois Environmental Protection Agency
LUST Section
1021 N. Grand Avenue, East

P.O. Box 19276
Springfield, IL 62794-9276

Indiana
Indiana Dept. of Environmental Management
Office of Environmental Response
UST Section
100 N. Senate Ave.
P. O. Box 7105
Indianapolis, IN 46206

Iowa
Iowa Dept. of Natural Resources
UST Section
Wallace State Office Bldg.
900 East Grand
Des Moines, IA 50319

Kansas
Kansas Dept. of Health and Environment
Bureau of Environmental Remediation
Storage Tank Section
Forbes Field, Bldg. 740
Topeka, KS 66620

Kentucky
Kentucky Division of Waste Management
UST Branch
14 Reilly Road
Frankfort, KY 40601

Louisiana
Louisiana Dept. of Environmental Quality
Office of Waste Services
UST Division
P.O. Box 82178
Baton Rouge, LA 70884-2178

Maine
Maine Dept. of Environmental Protection
Bur. of Remediation and Waste Management
Ray Bldg., Station #17
Augusta, ME 04333

Maryland
Maryland Dept. of Environment

Waste Management Administration.
Oil Control Program
2500 Broening Highway
Baltimore, MD 21224

Massachusetts
UST Contact:
Massachusetts Dept. of Revenue
Underground Storage Tank Program
200 Arlington St.
Chelsea, MA 02150
LUST Contact:
Massachusetts Dept. of Environmental Protection
One Winter St.
Boston, MA 02108

Michigan
Michigan Dept. of Environmental Quality
Storage Tank Division
P.O. Box 30157
Lansing, MI 48909-7657

Minnesota
Minnesota Pollution Control Agency
UST/LUST Program
520 Lafayette Rd. North
St. Paul, MN 55155-3898

Mississippi
Mississippi Dept. of Environmental Quality
Bureau of Pollution Control
UST Section
P.O. Box 10385
Jackson, MS 39289-0385

Missouri
Missouri Dept. of Natural Resources
Hazardous Waste Program
Tanks Section
P.O. Box 176
Jefferson City, MO 65102-0176

Montana
UST contact:
Montana Dept. of Environmental Quality

Waste Management Division
P. O. Box 200901
Helena, MT 59620-0901
LUST contact:
Dept. of Environmental Quality
Remediation Division
P.O. Box 200901
Helena, MT 59620-0901

Nebraska
UST Contact:
Nebraska State Fire Marshal
Flammable Liquid Storage
246 South 14th St.
Lincoln, NE 68508
LUST Contact:
Nebraska Dept. of Environmental Quality
LUST/ER Section
Box 98922
Lincoln, NE 68509-8922

Nevada
Nevada Dept. of Conservation and Natural Resources
Division of Environmental Protection
333 W. Nye Ln.
Carson City, NV 89706

New Hampshire
New Hampshire Dept. of Environmental Services
Bureau of Waste Management
P.O. Box 95
6 Hazen Dr.
Concord, NH 03302

New Jersey
New Jersey Dept. of Environmental Protection
Division of Responsible Party Site Remediation
P. O. Box 433
401 E. State St.
Trenton, NJ 08625-0433

New Mexico
New Mexico Environment Dept.
UST Bureau

P.O. Box 26110
Harold Runnels Bldg., Rm N2150
Santa Fe, NM 87504-0968

New York
New York Dept. of Environmental Conservation
Bulk Storage Section
50 Wolf Rd., Rm. 360
Albany, NY 12233-3750

North Carolina
Department of Environment and Natural Resources
Division of Waste Management
UST Section
2728 Capital Blvd.
Raleigh, NC 27604

North Dakota
North Dakota Dept. of Health
Division of Waste Management
UST Program
1200 Missouri Ave.
P.O. Box 5520
Bismarck, ND 58506-5520

Ohio
Ohio Dept. of Commerce
Bureau of UST Regulations
6606 Tussing Rd.
Reynoldsburg, OH 43068

Oklahoma
Oklahoma Corporation Commission
Petroleum Storage Tank Division
Jim Thorpe Bldg.
2101 N. Lincoln Blvd.
Oklahoma City, OK 73105

Oregon
Oregon Dept. of Environmental Quality
UST Program
811 SW Sixth Ave.
8th Floor
Portland, OR 97204

Pennsylvania
UST Contact:
Pennsylvania Dept. of Environmental Protection
Bureau of Watershed Conservation
Division of Storage Tanks
Rachel Carson State Office Bldg.
P.O. Box 8762
Harrisburg, PA 17105-8762
LUST Contact:
Pennsylvania Dept. of Environmental Protection
Bureau of Land Recycling and Waste Management
Land Recycling and Cleanup Program
Storage Tank Cleanup Section
Rachel Carson State Office Bldg.
P.O. Box 8471
Harrisburg, PA 17105-8471

Rhode Island
Dept. of Environmental Management
Office of Waste Management
UST Management
235 Promenade St.
Providence, RI 02908

South Carolina
South Carolina Dept. of Health and Environmental Control
UST Program
2600 Bull St.
Columbia, SC 29201

South Dakota
South Dakota Dept. of Environment and Natural Resources
UST Program
523 E. Capitol
Pierre, SD 57501

Tennessee
Tennessee Dept. of Environment and Conservation
Division of Underground Storage Tanks
4th Floor, L & C Tower
401 Church St.
Nashville, TN 37243-1541

Texas
Texas Natural Resources Conservation Commission
Registration and Evaluation Division

MC: 133
P.O. Box 13087
Austin, TX 78711-3087

Utah
Utah Dept. of Environmental Quality
Bureau of Solid and Hazardous Waste
Division of Environmental Response and Remediation
168 North, 1950 West
Salt Lake City, UT 84116

Vermont
Agency of Natural Resources
Division of Waste Management
Underground Storage Tank Section
103 S. Main St., West Building
Waterbury, VT 05671-0404

Virginia
Virginia Dept. of Environmental Quality
Underground Storage Tanks
P.O. Box 10009
Richmond, VA 23240

Washington
Washington Dept. of Ecology
Toxics Cleanup Program
P.O. Box 47600
Olympia, WA 98504-7600

West Virginia
UST Contact:
West Virginia Division of Environmental Protection
Compliance Assurance and Emergency Response Section
UST Program
1356 S. Hansford St.
Charleston, WV 25301-1401
LUST Contact:
Office of Environmental Remediation
LUST Program
1356 S. Hansford St.
Charleston, WV 25301-1401

Wisconsin
UST Contact:
Wisconsin Dept. of Commerce
Environmental and Regulatory Services Division

Bureau of Storage Tank Regulation
P.O. Box 7837
201 W. Washington Ave.
Madison, WI 53707-7837
LUST Contact:
Wisconsin Department of Natural Resources
Bureau of Remediation and Redevelopment
P.O. Box 7921
Madison, WI 53707-7921

Wyoming
Wyoming Dept. Of Environmental Quality
Water Quality Division
Herschler Bldg.
4th Floor West
122 W. 25th St.
Cheyenne, WY 82002

U.S. Territories

American Samoa Environmental Protection Agency
Office of the Governor
American Samoa Government
ATTN: UST/LUST Program
Pago Pago, AS 96799

Northern Mariana Islands
Division of Environmental Quality
Commonwealth of Northern Mariana Islands
P.O. Box 1304
3rd Floor, Morgens Bldg.
San Jose
Saipan, MP 96950

Guam
Guam Environmental Protection Agency
Air and Land Division
15-6101 Mariner Ave.
Tiyan, GU 96913

Puerto Rico
Environmental Quality Board
UST Control Program
P.O. Box 11488
Commonwealth of Puerto Rico
Santurce, PR 00910

Virgin Islands
Division of Environmental Protection
Dept. of Planning and Natural Resources
1118 Water Gut Homes
Christiansted
St. Croix, VI 00820-5065

12
Steel UST Technologies

Jim Wisuri
Prairie Oak Communications, Inc., River Forest, Illinois

I. NEW TECHNOLOGIES

During the 1960s and 1970s it became clear that the era of dominance for unprotected steel underground storage tanks was near an end, which led steel tank manufacturers to develop several technologies that would better serve UST customers. The tank technologies during the past three decades have reflected distinctions between unique approaches to providing corrosion protection and enhancing the long-term reliability of steel USTs. The technological approaches can be divided into several categories:

Cathodic protection
Composite
Jacketed
Thick urethane coated

The technologies, covering tanks fabricated by dozens of different manufacturers, have in many cases built national identities because they were backed by technical standards cited in federal and state regulations for USTs.

Technical standards—developed by organizations such as Underwriters Laboratories (UL) and Steel Tank Institute (STI)—enable, for example, a composite tank built by one company in California to have virtually identical characteristics to a composite tank fabricated by another manufacturer in Florida.

II. COMMON TANK TECHNOLOGY CHARACTERISTICS

In each category, a corrosion-protected steel UST built to Steel Tank Institute standards shares certain traits:

UL 58 primary tank
Dielectric corrosion barrier that protects the tank's external shell
Compatibility with a broad range of fuels and chemicals
Striker plates under each tank opening to combat internal corrosion
Capacities up to 50,000 gal
Options to customize compartments in the tank for multiproduct storage
30-year warranty against internal and external corrosion

III. CATHODICALLY PROTECTED

The most widely manufactured steel tank technology since 1969 has been the sti-P_3®, which features three levels of protection against corrosion—the use of sacrificial anodes, a durable coating, and electrical isolation between the tank and piping attachments. Built to conform with UL 58, UL 1746 and the sti-P_3 specification of Steel Tank Institute, the brand has been well known for the more than 50 manufacturers licensed to fabricate sti-P_3 in either single-wall or double-wall designs. STI also provides an international quality assurance/quality control program that supplements in-house quality efforts with random, unannounced, third-party inspections at licensees' fabrication plants.

The sti-P_3 technology features:

Unique, ongoing monitoring of corrosion protection
Double-wall designs with interstitial leak detection monitoring pipe included
An array of urethane, fiberglass-reinforced plastic (FRP) or coal-tar epoxy coating options [1]

IV. COMPOSITE

The origin of the composite tank actually predates sti-P_3. A few steel tank fabricators during the mid-1960s experimented with applying a thick layer of FRP to the exterior tank shell as a method of combating corrosion. In 1967, Steel Tank Institute introduced a new standard (STI-LIFE) for FRP-coated steel tanks. Tanks were actively built to the STI-LIFE standard for about five years. Manufacturers eventually replaced the standard with the more economical, production-friendly sti-P_3 tank.

However, because the marketplace resisted premium-priced USTs prior to the inception of federal and state regulatory programs, the composite tank languished until the 1980s without sparking widespread enthusiasm among tank buyers. During the past 15 years, its popularity has grown significantly among many UST specifiers and owners.

There are three primary brand names within the composite tank market: ACT-100®, Glasteel, and Plasteel. Each composite tank brand shares several characteristics. They are built to conform to UL 1746, Part II. They are available in single-wall, double-wall, and custom designs such as compartment tanks or oil-water separators. The abrasion-resistant FRP cladding is 100 mils on the external tank shell to prevent corrosion. The integrity of the cladding is verified both at the factory and prior to installation through holiday testing at 35,000 V. Double-wall tanks include a pipe to enable leak detection via interstitial monitoring. Composite tanks also feature the industry standard 30-year warranty [1].

A. ACT-100

This composite tank is best known for its signature method of corrosion protection—a durable 100 mils combination of isophthalic resin and chopped fiberglass that creates a hard and abrasion-resistant coating. Manufacture of such tanks is governed by the ACT-100 specification, which was initially developed by the now-defunct Association for Composite Tanks during the late 1980s and refined in recent years by STI. Besides the FRP coating, the specification calls for electrical isolation—typically accomplished with nylon bushings—between the tank and any piping attachments.

As it does with sti-P_3 and other tank technologies for which it grants licenses, STI provides an international quality control program that conducts random, unannounced inspections at manufacturers' facilities. STI initiated a standard for composite tanks (unrelated to STI-LIFE), and eventually merged the standard with the ACT-100 spec. ACT-100 tanks are fabricated and coated in a variety of colors, which have no technical effect on the performance of the FRP-and-resin corrosion barrier. The colors are simply a cosmetic marketing enhancement added by individual manufacturers.

B. Glasteel

Modern Welding Co. has established the Glasteel brand, which applies to composite tanks fabricated by any of the company's eight manufacturing locations in the United States and licensees in South America and Australia. Many Glasteel tanks also carry the ACT-100 label [2].

C. Plasteel

Plasteel is a trade name and trademark established by Joor Manufacturing Inc. and controlled today by Plasteel International Inc. Plasteel licensees can be found in North America, South America, Asia, and Europe. Joor obtained a UL listing for a corrosion-protected tank in 1980, which became known as a composite tank [3].

V. JACKETED

This is the newest and fastest-growing segment of steel UST technology. During the past 10 years, several unique designs have been developed to provide corrosion-protected secondary containment without employing two layers of steel. Jacketed tanks are built to conform with UL 1746, Part III.

A jacketed design employs a UL 58 primary tank encapsulated by a nonmetallic layer of protection. The secondary containment may be made of an FRP combination, a high-density polyethylene or a polymer—all of which provide a durable outer layer resistant to corrosion and compatible with a wide range of petroleum and chemical products. One of the most popular features is vacuum testing—holding 1 in. of Hg vacuum, which equals 2.036 psi—a typical method of showing that a jacketed tank is tight. Jacketed tanks also can be configured for multiple compartments. All of the jacketed tanks feature 30-year limited warranties.

A. Permatank®

Permatank is another underground storage tank technology licensed to STI members. Developed in 1987, it provides an FRP jacket as the secondary containment. Permatank ships from the factory with a vacuum drawn on the interstice to ensure the tank is tight when delivered. STI provides an international quality control program that conducts random, unannounced inspections at Permatank manufacturers' facilities [1].

B. Glasteel II

Glasteel II is a unique design developed by Modern Welding Co. It also employs protection to the primary tank with an outer wall of FRP [2].

C. Titan Tank

This technology, introduced in 1998, features an outer tank constructed of polyolefin. The tank is shipped with a built-in vacuum tightness system to assure the integrity of both the primary and secondary containments [4].

D. Total Containment

The Total Containment design relies on a high-density polyethylene (HDPE) design to enclose the steel inner tank [5].

E. Elutron

Elutron, which can be manufactured by the same companies that have rights to use the Plasteel name, also uses an FRP outer barrier to deliver secondary containment of the primary tank. The fiberglass outer shell is separated from the steel inner tank by an aluminum foil. Elutron was first developed in 1989 [3].

VI. THICK-FILM URETHANE COATING

One of the newest UST technologies—developed in the mid-1990s—features a thick-film urethane coating, which meets the requirements of UL 1746 and STI's ACT-100-U standards. Prior to the development of ACT-100-U, urethane-coated tanks required cathodic protection to meet national standards. However, the combination of FRP and urethane—in a thick-film coating—met UL and STI performance standards that certified the new technology as reliable.

Urethanes typically resist impact well and provide corrosion protection to the tank by totally isolating the steel from the soil.

ACT-100-U also is covered by STI's quality control inspection program.

REFERENCES

1. Steel Tank Institute. [Background posted on the World Wide Web.] Lake Zurich, IL. Retrieved October 24, 1998, from the World Wide Web. http://www.steeltank.com
2. Modern Welding Company. [Background posted on the World Wide Web.] Owensboro, KY. Retrieved October 24, 1998, from the World Wide Web. http://www.modweldco.com
3. Plasteel International, Inc. [Background posted on the World Wide Web.] Escondido, CA: retrieved October 24, 1998, from the World Wide Web. http://www.plasteel.com
4. Titan: One Tough Tank Product Literature. Utica, NY: Mohawk Metal Products Co., 1998.
5. Total Containment, Inc. [Background posted on the World Wide Web.] Oaks, PA, retrieved October 24, 1998, from the World Wide Web. http://www.totalcontainment.com

13
Underground Storage Tank Installation

Jim O'Day
O'Day Equipment, Inc., Fargo, North Dakota

I. INSTALLING IT RIGHT THE FIRST TIME

High-quality underground storage tank installations result from hiring reputable, qualified contractors. But tank owners, specifiers, and managers cannot rely solely on the contractor. End users and consultants should educate themselves, at a minimum, by reading several documents:

1. PEI RP 100 (Recommended Practices for Installation of Underground Liquid Storage Systems Tanks for Flammable and Combustible Liquids), published by the Petroleum Equipment Institute
2. NFPA 30 (Flammable and Combustible Liquids Code) and NFPA 30A (Automotive and Marine Service Station Code), published by the National Fire Protection Association
3. The tank manufacturer's installation instructions

The next essential step is to follow up with the manufacturer or installer—prior to the start of work—with any questions about the tank or the installation process.

Warranty and insurance claims made on tank leakage can often be traced to poor handling and installation. So, by striving for installation excellence, end users and contractors can save all parties concerned needless headaches. It truly is less work—and much less expensive—to do it right the first time.

II. INSTALLING STEEL TANK TECHNOLOGIES

Today's installers must be knowledgeable about a variety of steel tank technologies—from cathodically protected to composite to jacketed designs. This chapter will spotlight certain aspects of the installation process that are important to each of the steel technologies. However, there are significant distinctions between tank designs for which installers must account. To ensure that each tank receives the best possible installation, it's critical for contractors to be intimately acquainted with the differences between steel UST technologies—and their unique needs for proper installation.

For instance, performing a 35,000 V holiday test on a urethane-coated sti-P_3 tank would raise questions about an installer's familiarity with proper procedures. (Such a holiday test is most appropriate for a composite tank that employs a thick laminate of Fiberglas-reinforced plastic [FRP]).

Some steel USTs shall be operated at ambient temperatures only. Check with the tank manufacturer prior to delivery on whether products such as heated oil may be stored in the tank.

III. TANK SYSTEM SITE PREPARATION

Planning should begin with a site evaluation—and some worst-case scenarios. What if something occurs—several days of rain, discovery of a prior petroleum leak at the proposed tank site, or an excavation cave-in that injured a worker—that would add days (or weeks) to the installation job?

If something happened, the job site may have to accommodate the storage of an underground storage tank and perhaps piping, backfill materials, and installation equipment—without disrupting the tank owner's business. At an industrial facility sited on several hundred acres, such a storage question is a relatively minor issue. At a suburban convenience store location on a small lot, the question can mean a significant loss of drive-in business.

The American Petroleum Institute (API) in its Publication 1615 (Installation of Underground Petroleum Product Storage Systems) suggests the importance of a site analysis prior to installation. Besides finding space to take care of equipment storage at the job site, the analysis could include:

- Depth of the water table
- Location of underground and aboveground utility structures, and foundations for nearby buildings
- Potential presence of contaminated soil
- Soil resistivity
- Soil acidity
- Moisture content and overall stability of the soil

Tank manufacturer's recommendations on how much distance should separate each tank in the excavation
Probable angle of slope for excavation walls
Required depth of the excavation
Placement of tank-anchoring devices (1)

By anticipating the source of possible delays, the UST project team can minimize, or eliminate, needless extra costs.

IV. EXCAVATION AND BEDDING

The excavation must be free from any hard or sharp materials that may cause damage to the steel UST. The contractor shall take care during installation to prevent the introduction of foreign matter into the excavation or backfill (2).

Bedding and backfill materials fulfill three key functions:

Providing a secure foundation and support to the tank structure
Isolating the tank from other materials that could damage the tank shell or coating
Providing a firm base for paving over the tank system (3)

Steel tank installation instructions also call for bedding and backfill material to be clean, homogeneous, granular material made of clean sand, pea gravel, No. 8 crushed stone (No. 8 coarse aggregate per American Society for Testing and Materials ASTM-D448, ½-in. maximum size), or equivalent.

The excavation's bottom must be covered with a minimum of 2 ft of bedding (609.6 mm), suitably graded and leveled. Some manufacturers may allow less backfill, but the industry standard is 2 ft. The excavation shall extend a distance of at least 2 ft (609.6 mm) around the perimeter of the tank providing sufficient clearance.

When a concrete slab is required for anchoring, the tank shall not be placed directly on the pad. Bedding material at least 12 in. (304.8 mm) deep must be spread evenly over the entire pad to separate the tank from the concrete. Some tank manufacturers may allow slightly less.

In a tidal area, the tank-bedding material shall be crushed stone or pea gravel. Sand backfill may be used only if measures are taken to prevent the washout of sand and the infiltration of native soil into the backfill during the design life of the system (4).

V. JOB SITE AIR TEST

This is an extremely effective means of assuring the integrity of the underground storage tank prior to installation. However, this test does not apply to jacketed tanks that are shipped from the factory with a vacuum on the interstitial space.

Steel tank manufacturers' installation instructions for composite, thick-film urethane and cathodically protected tanks say that temporary plugs and thread protectors installed in the fittings by the manufacturer shall be removed. Following that, the contractor must apply compatible, nonhardening pipe sealant to internal bushing threads (Fig. 1). (Nylon bushings provide electrical isolation between the tank and any metallic attachments routed through the fittings.) Permanent metal plugs shall be installed at all unused openings (5).

Isolation bushings shall not be removed from the unused openings. Likewise, plugs used to temporarily seal the tank for the aboveground air test, but later removed for pipe installation, shall not be overtightened. Installers must avoid cross threading, or other damage to nylon bushings, when replacing plugs or installing required tank piping.

Test pressure shall be maintained at, but not exceed, 5 psi (34.47 kpa) while a soap solution covers the tank (Fig. 2). Installers particularly need to check pipe connections and welds for the presence of bubbles that indicate a leak.

Dual-wall steel tanks require different air pressure testing procedures. The two most common tests are a vacuum on the interstitial space, or an air test on the primary tank in which the pressure is bled carefully into the interstitial space. Consult with the tank's fabricator for air test recommendations. A high-pressure air line should never be connected directly to the interstitial monitoring port. Do not apply a vacuum to the primary tank or a single-wall tank. PEI RP 100 (Recommended Practices for the Installation of Underground Liquid Storage Systems) also provides guidelines.

The need to take necessary safety precautions during air tests cannot be overemphasized. USTs must be attended while under pressure. While monitoring the air pressure test, avoid standing at the head of the tank. An air pressure relief valve is required (2).

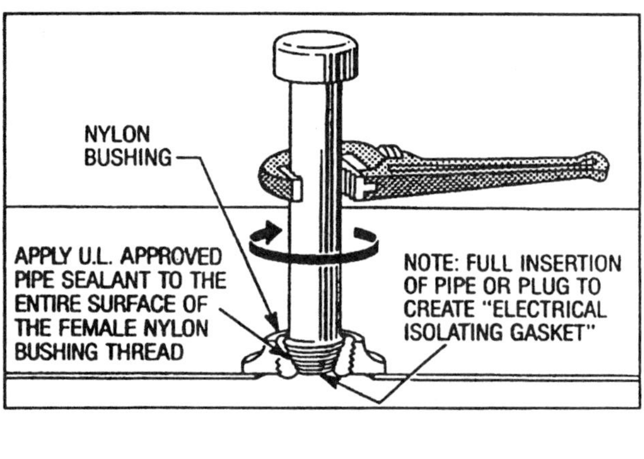

Figure 1

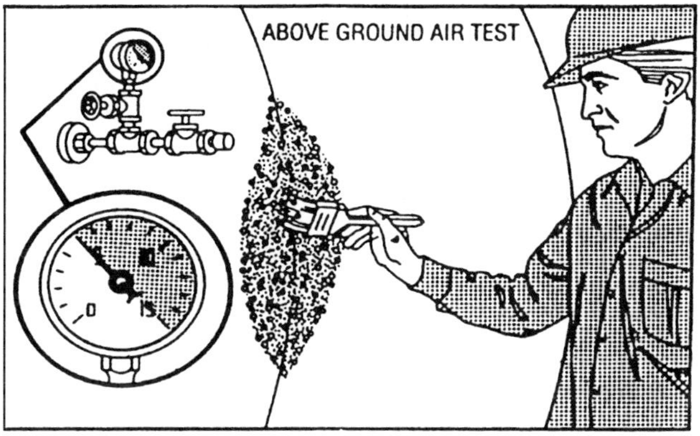

Figure 2

VI. TANK INSPECTION

Some jacketed and thick-film urethane tanks are shipped with protective isolation board attached. This board shall remain on the tank and not be removed.

For all steel USTs, before placing the tank in the excavation, any dirt clods and similar foreign matter shall be cleaned from the surface of the tank.

Before moving the tank into the excavation, it is critical to visually inspect the structure for damage. The installer must particularly pay attention to areas where coating or cladding has been gouged or abraded, then mark all areas that appear damaged for repair. All holidays, damaged coating or cladding, and/or exposed steel surfaces must be touched up using a kit furnished by the tank supplier. Follow manufacturer's instructions for mixing and application of resin. Holidays repaired on FRP-coated tanks may require a spark test to demonstrate that no steel has been left exposed. The installation project team should check with the manufacturer and authorities having jurisdiction (AHJ) on spark test requirements.

VII. TANK HANDLING

Equipment to move the tank shall be of an adequate size to lift and lower the tank without dragging or dropping. Careful handling will prevent damage to the tank or the coating.

The tank shall be carefully lifted and lowered off of the truck and into the excavation hole by use of cables or chains of adequate length attached to the lifting lugs provided. A spreader bar should be used when necessary. Chains or wire rope slings must not be used around the tank itself (4).

PEI RP 100 says a UST may be subjected to minimal movement prior to installation for tasks related to inspection and testing. During the soaping of a tank for the air test, it may be necessary to roll the UST slightly to make sure the shell's entire surface is soaped and inspected.

If a tank will be stored at an installation site for a few days, the structure should be chocked firmly to prevent uncontrolled rolling. Old automobile tires are ideal for chocking (6).

VIII. ANCHORING

High-water tables or partially flooded excavation sites exert significant buoyant forces on tanks. Buoyant forces are partially resisted by the weight of the tank, the backfill, and the pavement atop the tank.

If substantial amounts of water enter an excavation while new tanks are empty, it may cause the storage system to float or change position. To prevent this problem, it is common practice to fill the tanks with liquid ballast as soon as it's feasible. Two types of ballast are most often used: water, or the product that the tank ultimately will store (7).

Each type of ballast poses unique challenges. The installation project team should develop an approach on a case-by-case basis that will pose the fewest problems for each project.

Additional buoyant restraint, when required, shall be obtained by using properly designed hold-down straps in conjunction with concrete hold-down pads or deadman anchors. The use of steel cable or round bar as hold-down straps is prohibited.

Installers of hold-down straps shall ensure that the straps are separated from the tank by a pad made of inert insulating di-electric material (Fig. 3). The insulating material shall be wider than the hold-down straps, which will prevent direct contact between the straps and the tank shell (2).

IX. BACKFILL

Homogeneous backfill material similar to the bedding material shall be carefully placed around the entire tank to create a uniform homogeneous environment. Installers must take care to avoid damaging the tank's coating or cladding, especially where tamping is required (5).

Installing and tamping backfill along the bottom sides of the tank will ensure that the tank is fully and evenly supported around the bottom quadrant (Fig. 4).

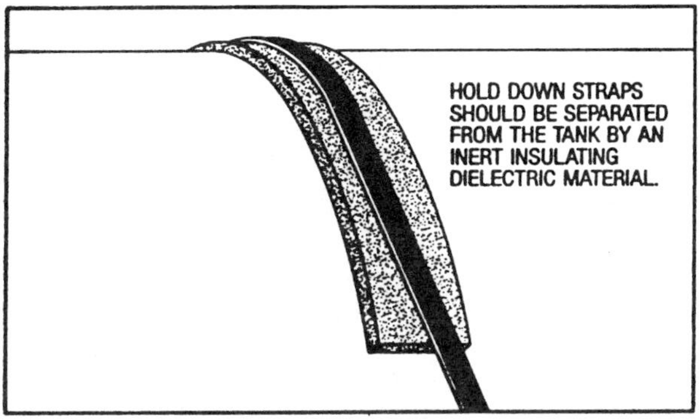

Figure 3

X. ELECTRICAL ISOLATION

For thick-film urethane, composite or cathodically protected USTs, all openings shall be isolated in which electrical accessories will be installed—such as a submersible pump, monitoring equipment, and all other grounded electrical devices.

Prior to backfilling over top of tank, all openings must be visually inspected to assure that the dielectric isolation (nylon bushings or flange gaskets), where furnished, remain in place. Where flanged openings have been used, isolation of the flange gaskets must be confirmed with a continuity tester (Fig. 5). The installation contractor should make contact with the inside of the tank—or in some cases, at

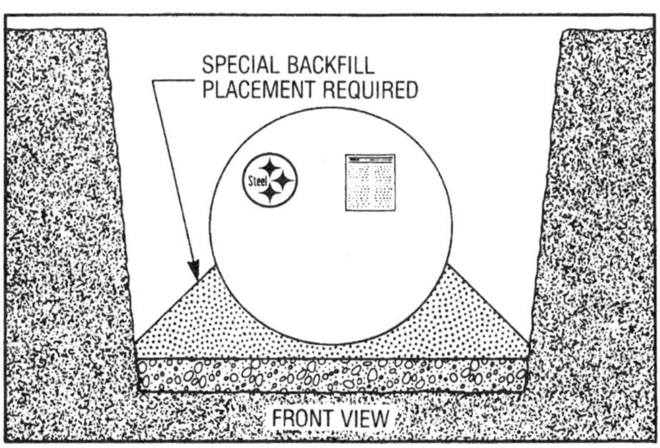

Figure 4

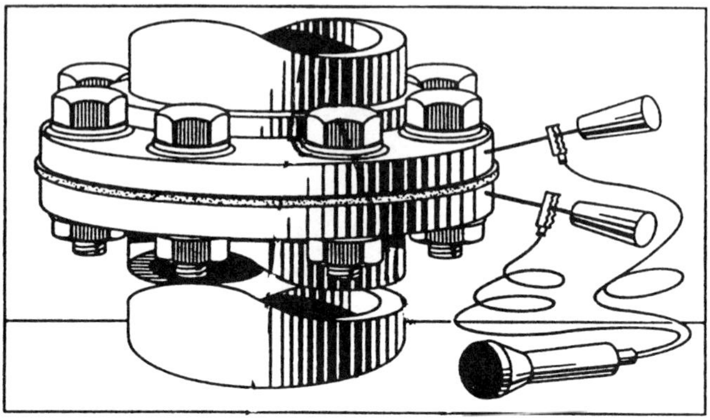

Figure 5

the lift lug, if steel is exposed, prior to application of a lift-lug cover. No current shall pass through the factory-installed flanged gaskets. Isolation of the fittings is required to assure tank integrity.

If the tank is to be installed in the vicinity of an impressed current cathodic protection system, consult with the tank manufacturer for additional instructions.

XI. FINAL AIR TEST

Tank piping must be installed using compatible, nonhardening sealant, taking care not to cross-thread or damage the nylon bushing. Torque of 400 to 1000 ftlb (542.3 to 1355.8 N-m) may be required to fully insert the pipe (4).

Where air or hydrostatic testing is required after installation, the pressure applied must not exceed 5 psi (34.47 kpa) as measured at the top of the tank. A soap solution shall be applied around pipe connections during the air test.

XII. SEALING OF PIPE CONNECTIONS, LIFTING LUGS, AND REPAIRS

During the installation process, steel can become exposed at the lift lug due to handling of the tank. Steel may also be exposed at the interface between the steel tank and the electrical isolation fitting. The installation contractor must cover these areas, along with all other exposed steel surfaces, by using the touchup kit supplied by the manufacturer. After application, the installer shall verify that all repaired areas have cured (yielding adequate material hardness and solidification) prior to backfill. Normal cure time may vary, in particular, depending on weather conditions.

The installer must clean any areas to be repaired through removal of surface rust, dirt, contaminants, and disbonded coating or cladding. The coating or cladding surrounding all holidays, flaw areas, and/or exposed steel should be surface prepared by using a coarse grit sandpaper or grinder. (Refer to the Steel Structures Painting Council SSPC SP-2 [Hand Tool Cleaning] or SP-3 [Power Tool Cleaning] for additional guidance.) This process should remove all glossiness from the surface surrounding the repair area within 6 in. of the holiday (2).

After an air test has established tightness, tank fittings shall be covered with a coat of the repair material prior to backfill. Areas to be coated shall include the entire plugs on unused fittings.

XIII. FINAL BACKFILL

The installer must carefully deposit homogeneous backfill around the tank and to a suitable depth—usually at least 3 ft (914.4 mm) over the tank (Fig. 6). (See PEI RP 100, NFPA 30, and NFPA 30A and state or local codes for minimum depth of cover required.)

If coating or cladding damage occurs after holiday testing or during the backfill operation, repairs shall be made in accordance with the instructions in the previous section.

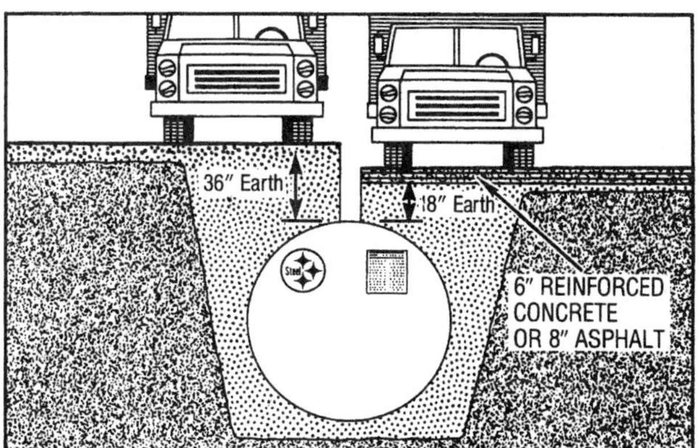

Figure 6

XIV. MAINTENANCE

The primary tank should be inspected monthly for the presence of water. Inspection should take place at the lowest possible points inside the primary tank. Remove any water found. Water and sediment in fuel can cause plugging of filters. Also, bacterial growth originating from the fuel can cause filters to plug and the corrosion of tanks and lines.

A report by the U.S. Department of Energy, BNL-48406, provides information on methods to test for and remove water, test for bacterial presence in fuel, tank cleaning, and fuel additives. Failure to remove water from the tank may void your warranty.

Some cathodically protected tanks will require cathodic protection monitoring. See Chapter 15 for more information on such monitoring.

REFERENCES

1. Excavating and trenching. Petroleum Equip Technol 1(6):23–24, 1996.
2. ACT-100-U Installation Instructions, R971-98 recommended practice. Lake Zurich, IL: Steel Tank Institute, 1998.
3. Backfill—selection, placement and compaction. Petroleum Equip Technol 1(3):29, 1996.
4. sti-P_3 Installation Instructions, R821-98 recommended practice. Lake Zurich, IL: Steel Tank Institute, 1998.
5. ACT-100 Installation Instructions, R913-98 recommended practice. Lake Zurich, IL: Steel Tank Institute, 1998.
6. Material handling on a tank project. Petroleum Equip Technol 1(1):23–24, 1996.
7. Backfill—selection, placement and compaction. Petroleum Equip Technol 1(3):32, 1996.

14
UST Performance

Brian C. Donovan
Containment Advisors, Inc., Lake Zurich, Illinois

I. INTRODUCTION

As underground storage tank (UST) technologies have come and gone, the recurring question has been: What works?

Since the early 1980s, and the congressional passage of Public Law 98-616, the tank performance issue has been a focal point for environmental solutions. Since the publication of the U.S. Environmental Protection Agency's (EPA) final UST regulation in 1988, the question of what works has centered on a variety of tank technologies with unique features that protect groundwater supplies and virtually eliminate the prospect of a leak. To gain perspective on the effectiveness of today's newest technologies, it helps to understand the efforts made during the 1980s and early 1990s to evaluate UST performance.

Looking back, the first thing to remember is that the performance problem was defined as an underground storage tank leak. Thousands of steel tanks—fabricated with ineffective or nonexistent corrosion protection—were leaking in communities and generating headlines across the nation. The tanks were the largest, most visible components of the storage system—and many were suffering from corrosion.

That steel buried in a moisture-rich environment would corrode shouldn't have been big news. Most school kids learn in science class that if you leave a steel nail in a jar of water overnight, a rusty residue will cloud the container in the morning. Corrosion is time sensitive, a naturally occurring process during which steel releases energy as it attempts to revert to iron ore. Bare steel tanks were leaking because they had outlived their useful life.

So, as environmental officials and legislators looked for solutions to polluted wells and waterways, the tank was defined as the culprit. During the last decade, the perception of the "tank problem" has slowly evolved into a "tank system problem." This has led to a greater emphasis on tank system management, which includes significant attention to preventing product releases from piping that could contaminate soil or water supplies.

The tank industry had known for years that there should be a better way to build USTs and prolong their life without sacrificing other important elements in tank storage. The fiberglass-reinforced plastic (FRP) tank debuted during the mid-1960s, as did a steel tank coated with FRP and other innovations designed to create a barrier between corrosive soil and a steel tank's external surface. The sti-$P_3^{®}$ tank—featuring cathodic protection, a durable coating, and electrical isolation between the tank and piping—was launched in 1969.

But in each case, the technologies were accepted slowly due to marketplace resistance. Petroleum marketers and other tank owners balked at the premium prices commanded by the eco-friendly designs—especially when they faced no legislative or regulatory mandates to use the new, enhanced tanks.

However, after Congress in 1984 approved Public Law 98-616—reacting to a barrage of media attention on UST systems that had leaked and caused numerous local problems—the marketplace resistance softened. And, after the federal EPA in 1985 issued an Interim Prohibition against the use of unprotected steel tanks, the tank-buying market truly began to pay attention when tank manufacturers offered corrosion-resistant designs.

As it started to develop a regulatory program, EPA in the mid-1980s was trying to understand and document the scope of the tank performance issue. The agency commissioned two major studies that shed light on what the new UST regulations would have to address. The first, released in 1986, was a 50-state effort, the Summary of State Reports on Releases from Underground Storage Tanks. The second, published more than two years later, examined the experience of one county as it documented old tanks removed from service. Instead of anecdotal reporting on tank performance (or failure), the two studies broke ground in documenting the dimension of the tank-leak issue.

II. 50-STATE STUDY (AUGUST 1986)

This was EPA's first attempt after congressional enactment in 1984 of Public Law 98-616, which amended the Resource Conservation and Recovery Act (RCRA), to learn more about the true scope of the tank leak problem. While the bulk of press coverage historically had blamed bare steel tanks for leaks, the 50-state study showed that one in five leak incidents involved FRP tanks. Eighty-one percent of incidents involved unprotected steel tanks, 19 percent FRP, according to the report [1].

Beyond that, the report was the first to characterize the "tank leak" problem as a system that needed correction. "The most frequent documented location of release is through the tank into the surrounding soil (43 to 58 percent)," the study said. "Piping accounts for between 20 and 35 percent of the reported release locations; overfills/spills account for 15 percent of those reported; and the remainder are due to pump and miscellaneous causes." The report showed that a comprehensive solution must address, the tank, piping, other fluid-containing components, and tank system management.

The 50-state survey attempted to associate tank age with failure rate. In retrospect, this approach reflected the relative lack of sophistication in the mid-1980s UST market. Though corrosion-protected tank technologies had been available for about 20 years, the EPA report drew no tank age distinctions between unprotected and protected tanks. The report did acknowledge that corrosion was the primary source of failure for steel tanks—and structural failure the leading cause of FRP tank leaks.

Based on the national evidence, the report concluded that "large tanks are as likely to leak as medium or small tanks"—a statement contradicted two years later by the findings of a study that examined 500 tanks removed from service in New York City's easternmost suburbs.

III. SUFFOLK COUNTY 500-TANK STUDY (NOVEMBER 1988)

The localized Tank Corrosion Study conducted in Suffolk County, New York—the eastern two-thirds of Long Island, and home to many underground tanks subjected to saltwater conditions—opened eyes in the industry with its six major conclusions [2]:

- Size is more important than age in predicting tank failure.
- In general, small tanks are much more likely to perforate than large tanks due to the thinner walls found in smaller tanks.
- Compared to external corrosion, internal corrosion is insignificant.
- Fuel oil tanks are just as susceptible to perforation as gasoline tanks of the same size.
- Existing tanks are in worse shape than is demonstrated by testing.
- Tanks do not always leak immediately upon perforation.

Beyond the report's major conclusions, the body of the work provided findings that confirmed earlier suppositions, while unveiling new evidence in several areas.

- Massive UST corrosion problems on bare steel tanks
- Some evidence of bare steel tanks withstanding corrosive conditions: only 58 percent of perforated tanks showed evidence of leakage—or 16.6 percent of the total sample

No evidence of problems with cathodically protected steel tanks

"The smaller tanks are more susceptible to perforation because they are made of thinner material, while the larger tanks are not as likely to fail because they are made with thicker material," the report said, recognizing the corrosion allowance built into steel tanks for decades by manufacturers. "Tanks 5,000 gallons and smaller accounted for 99.3 percent of all perforated tanks ... only one tank larger than 5,000 gallons had perforations" (Table 1).

At the same time, the report dashed the contentions that bare steel tanks were due to fail after a few years. "The range of ages for perforated tanks was between

Table 1 Tanks Removed from Service by Capacity

Volume (gal)	All tanks	Perforated tanks	Percentage perforated
175	1	0	0
185	1	1	100
275	18	5	27.8
315	1	1	100
500	2	1	50
550	58	13	22.4
575	1	0	0
1,000	64	33	51.6
1,100	1	0	0
1,500	8	2	25
2,000	73	35	47.9
2,500	5	0	0
3,000	59	21	35.6
3,500	1	0	0
4,000	65	25	38.5
5,000	34	5	14.7
6,000	12	0	0
7,500	5	0	0
8,000	12	0	0
10,000	51	0	0
12,000	3	1	33.3
15,000	8	0	0
20,000	5	0	0
25,000	8	0	0
30,000	2	0	0
50,000	2	0	0

Source: Ref. 2.

8 and 70 years, with perforations scattered throughout [the range]," the report said (see Table 2).

The message of the two EPA-funded studies was clear: the Interim Prohibition was on target. There would be no going back to an era in which corrosion protection was given lip service.

IV. TILLINGHAST REPORT (1993)

During the 1970s and 1980s, the cathodically protected sti-P_3 technology became the leading steel underground storage tank built in America. Nearly 60 manufacturers across the nation were qualified and licensed to fabricate sti-P_3 tanks, which were distinguished in the marketplace by anodes most commonly attached to the heads of the tanks. The anodes provided low levels of electrical current that would prevent corrosion underground if the tank's coating should be slightly flawed, or damaged during installation.

When the EPA's final regulations were released in 1988, the sti-P_3 tank was put at a competitive disadvantage. Owners of single-wall sti-P_3 tanks were required to conduct cathodic protection monitoring within six months of installation and at least once every three years to ensure that the preengineered corrosion protection was continuing to operate as designed. EPA exempted double-wall sti-P_3 tanks from the cathodic protection-monitoring requirement.

To address the regulatory mandate, Steel Tank Institute in 1988 developed the Watchdog program, which provided cathodic protection monitoring services to owners of sti-P_3 tanks. Watchdog technicians gathered the readings using a voltmeter and a reference electrode, and stored the information in hand-held computers that eventually transferred the test results to a central database. Watchdog program participants received reports for their files of cathodic protection readings.

By the early 1990s, marketplace misperceptions that cathodic protection monitoring was complex, onerous, and expensive had diminished demand for sti-P_3 technology. Steel Tank Institute, on behalf of its member manufacturers, contracted with Tillinghast [3], a prominent risk management–consulting firm with experience in the underground storage industry, to study the performance of sti-P_3 tanks.

"Few low readings (on sti-P_3 tanks) for cathodic protection system testing were reported from either Watchdog participants or major oil companies who maintain their own monitoring programs," the Tillinghast report said.

> Given: 1. technical uncertainty over the corrosion monitoring requirement; 2. the very low overall incidence rate for external corrosion; and 3. No reported instances of external corrosion for sti-P_3 tanks that have been properly fabricated, transported and installed, it appears that the frequency of monitoring is a topic that should be revisited [by EPA]. Regulators, tank own-

Table 2 Age of Tanks Removed from Service

Age in years	All tanks	Perforated tanks	Percentage perforated
70	1	1	100
60	1	0	0
57	1	1	100
55	1	1	100
50	1	1	100
48	5	3	60
47	1	0	0
46	1	0	0
44	12	1	8.3
43	7	3	42.9
41	2	0	0
40	4	1	25
37	2	1	50
36	1	0	0
35	3	0	0
34	1	0	0
33	2	1	50
32	6	0	0
31	5	0	0
30	14	7	50
29	3	0	0
28	17	3	17.6
27	14	6	42.9
26	22	16	72.7
25	10	3	30
24	9	1	11.1
23	26	8	30.8
22	10	1	10
21	9	3	33.3
20	23	6	26.1
19	16	9	56.3
18	15	7	46.7
17	25	9	36
16	28	13	46.4
15	18	4	22.2
14	16	6	37.5
13	19	2	10.5
12	8	1	12.5
11	7	3	42.9
10	30	2	6.7
9	7	0	0
8	3	1	33.3
7	5	0	0

Table 2 Continued

Age in years	All tanks	Perforated tanks	Percentage perforated
3	1	0	0
2	1	0	0
Unknown	87	18	20.7
Total	500	143	

Source: Ref. 2.

ers and installers who responded to questions on corrosion monitoring frequency offered a wide range of opinions with no technical justification.

The survey of tank owners covered over 3,000 sti-P_3 tanks, and the survey of tank installers covered over 5,000 sti-P_3 installations. Of the 8,000+ sti-P_3 tanks, three instances of external corrosion were reported, representing a frequency of 0.04 percent. Only one of the three instances involved a product release.

Tillinghast reported that environmental and products liability insurance claims on sti-P_3 tanks "were few." Only four out of 103 closed claims (3.9 percent) involved damages where money was paid out. Many claims only incurred investigative and administrative costs.

The corrosion-monitoring requirement and the technical basis on which this mandate is based are not understood by most tank owners, installers, or regulators, Tillinghast contended. Many tank owners were simply unaware of the monitoring requirement. Similarly, enforcement of the cathodic protection monitoring requirement was not a high priority for EPA regional and state regulators, the report said.

> It is clear from our interviews with regulators that they maintain a healthy skepticism towards sti-P_3 tanks. Several state and EPA regional officials were forming their opinions on experiences with bare steel tanks. They have little or no technical data to support their skepticism for sti-P_3 tanks, but several regulators believe there have been or will be releases resulting from external corrosion. At the same time, regulators who have witnessed the removal of sti-P_3 tanks reported that the tanks and sacrificial anodes were in excellent condition when the tanks were removed.

Despite the evidence to the contrary, EPA left intact the cathodic protection monitoring requirement for single-wall sti-P_3 tanks.

But the Tillinghast report provided third-party documentation on how far the UST industry had traveled—from the days of unprotected steel tanks—in developing cost-effective, highly reliable, nearly perfect tank designs.

Nevertheless, other underground storage tank performance questions arose during the 1990s. Can fiberglass-reinforced plastic USTs reliably provide long-term structural strength and compatibly store alternative fuels with high alcohol content? Can the FRP cladding on a composite tank provide enough structural strength to offset the use of thinner-gage steel? At what thickness does steel truly provide structural integrity? (Chapter 8 provides more detail on studies that show how the most basic element of steel tank design was challenged by a late 1980s technological approach.)

V. CONCLUSION

In the late 1990s, any number of UST technologies are on the market bearing Underwriters Laboratories listings and backed by various quality control efforts, warranties, and leak detection systems. In less than two decades, the UST performance question has all but vanished. Leaks from new technology UST designs are rare, rather than commonplace—a performance turnaround with environmental significance for future generations.

REFERENCES

1. Summary of State Reports on Releases from Underground Storage Tanks. Washington: Versar, Inc. 1986.
2. JH Pim, JM Searing. Final Report: Tank Corrosion Study. Suffolk County, N.Y. Farmingville, NY: Department of Health Services, 1988.
3. Tillinghast, a Towers Perrin Company. Evaluation of the Potential for External Corrosion and Review of Cathodic Protection Monitoring Associated with sti-P3 Underground Storage Tanks. Arlington, VA: 1993.

15

Corrosion and Cathodic Protection on Underground Storage Tanks

Taylor Leon
Steel Tank Institute, Lake Zurich, Illinois

I. CORROSION AND CATHODIC PROTECTION

The act of corrosion is a natural process for many materials. The rate at which metals corrode depends on their innate energy levels. When electrically connected, metals with high energy levels are used as anodes to protect metals at lower energy levels. Examples of high–energy level metals are magnesium and zinc. A lower–energy level metal such as steel becomes a cathode when it is coupled with a high–energy level metal. The protective corrosion cell becomes active when both metals are placed in the same electrolyte and are electrically connected via a metallic path. In practical application, a protective corrosion cell is known as a cathodic protection system.

Cathodic protection is a widely used method of preventing the corrosion of steel in an electrolyte. It has been effectively used for more than a century in the marine industry and for over 50 years in the refining and pipeline industries. The protective relationship between anodes and cathodes is effectively employed in many consumer industries on products such as water heaters or boat-engine propellers. The electrochemical process produced by cathodic protection utilizes the basic elements required for any corrosion cell. Those elements are an anode, a cathode, electrolyte, and a metallic path.

The anode is the part of the corrosion cell from which positive metal ions flow into the electrolyte. It is the site on the metal structure where corrosion and metal loss occurs. This action produces a measurable DC voltage.

The cathode is the part of the corrosion cell that accepts the metal ions flowing from the anode when the two are electrically connected. (A cathode can easily become an anode when it is not electrically connected to an anode.)

For example, a copper line is connected to a steel fitting, and both are connected to magnesium. In this scenario the magnesium is anodic with respect to the copper and steel because it is a more active metal. Therefore, the copper/steel couple is cathodic and current flows (through the corrosion process) from the magnesium to the copper/steel couple. If the magnesium is disconnected from the copper/steel couple, the steel becomes anodic with respect to the copper. This results in the copper being cathodic and as a result, current flows from the steel to the copper. This would result in metal loss for the steel—or corrosion. The anode-cathode relationship is easily recognized using the galvanic series.

The electrolyte is an environment, typically soil or water, which allows current to flow. The degree of current flow is a function of conductivity or resistance of the electrolyte.

A cathodic protection system is typically monitored according to guidelines established by NACE International, an association of corrosion engineers. Cathodic protection monitoring provides a means to demonstrate whether a metallic structure is protected against corrosion. The cathodic protection reading may be based on the NACE International standard of $-.85$ V. Cathodic protection readings equal to or more negative than $-.85$ V indicate that the cathodic protection system is sufficiently controlling the corrosion of steel. Another guideline established by NACE International is the 100-mV polarization criteria, which will be discussed later in this chapter.

The U.S. Environmental Protection Agency (EPA) requires that some cathodically protected underground storage tanks be monitored to ensure the effectiveness of the cathodic protection system. The monitoring frequency, according to the federal regulation, is within six months after installation and every three years thereafter.

The sti-P_3® tank, which is built by Steel Tank Institute licensees, is primarily affected by the cathodic protection monitoring requirement. However, EPA has issued two interpretations that affect both sti-P_3 and other tanks that employ anodes as a protection against corrosion. The first pertains to double wall sti-P_3 tanks with operational interstitial monitoring devices, capable of detecting breaches through the primary or secondary wall, on a monthly basis. The second interpretation affects composite ACT-100® or ACT- 100-U® tanks with anodes installed per the sti-P_3 specification and anode bars properly coated. Both systems must be tested for proper cathodic protection operation at the time of installation and then, whenever additional construction or maintenance takes place around the tank.

Qualified individuals are supposed to perform the monitoring. Qualification—based on education, skill, and experience—is often obtained through certification by a national trade association. Individual states can require monitoring at a more frequent interval and may allow qualification through other means such as successful completion of a recognized course of study.

When monitoring cathodic protection systems several techniques can be used to perform the test. The cathodic protection test is the primary method used to assure a cathodic protection system is working properly. The test is performed to measure an electrical potential at a buried metallic structure that is believed to be under the influence of cathodic protection. The most popular techniques for structure-to-soil potential readings will be discussed.

II. EQUIPMENT

A. DC Voltmeter

The typical meter used for the cathodic protection test is a digital type with at least 10 megohm internal-circuit resistance (impedance). The meter must be capable of measuring levels of DC voltage below 2 V.

B. Reference Electrode

The reference electrode, often referred to as a half-cell, may be one of several types. The most commonly used for buried (or freshwater submerged) application is the copper/copper sulfate electrode (CSE). The reference electrode consists of a hollow plastic cylinder with a porous tip on one end. The other end is plugged with a cap that has a copper rod extending through it. The body of the cell contains a solution consisting of copper sulfate crystals (half full) and distilled water (two-thirds full). The solution of copper sulfate is maintained in a saturated state (crystals visible).

C. Test Leads and Electrical Wire

Test leads should be available for connection from the DC voltmeter to the structure under test and the CSE. Additional electrical wire may be required to perform troubleshooting and other testing. The wire should be insulated 12 gauge. Clip connectors must be installed on the ends of any wire to be used for additional test connections.

III. MAINTENANCE

Although a minimal amount of equipment is needed to perform a cathodic protection test, it is imperative that the equipment be the correct type and in good working condition. The digital meter requires little maintenance. The primary maintenance item is the battery. Most digital voltmeters have a visible "low battery" indicator that signals the time to replace the power source. It is advisable to change the battery as soon as possible when the warning light appears because the power level may directly affect the accuracy of the voltmeter's reading.

The reference electrode also requires little maintenance. The level of distilled water should always be maintained near the two-thirds-full level. The solution is active as long as crystals are visible. Eventually, the copper rod will require removal and cleaning. The cleaning is accomplished by removing patina (green stain) from the rod using sandpaper. Do not use metallic grit paper or steel wool. The use of a metallic abrasive on the rod will embed metallic traces in the rod, which can cause inaccurate performance.

The porous tip will wear down after repeated use. As the tip wears down, the surface area of the tip contacting the soil decreases. When the surface area of the tip is reduced drastically there is a greater potential for an inaccurate reading. The electrode is designed for optimum performance when all of the components are maintained in a nearly new condition.

The metallic connector on each end of the test lead or cable is less effective when oxidation is present. The added resistance caused by the oxide can affect the reading accuracy. Connectors should be maintained in a clean condition.

IV. METER CONNECTIONS

There are two different approaches regarding the way the meter is connected between the structure to be tested and the reference electrode.

The first approach uses a digital multi-meter (Fig. 1). When employing this equipment, connect the positive meter lead (typically red) to the structure under

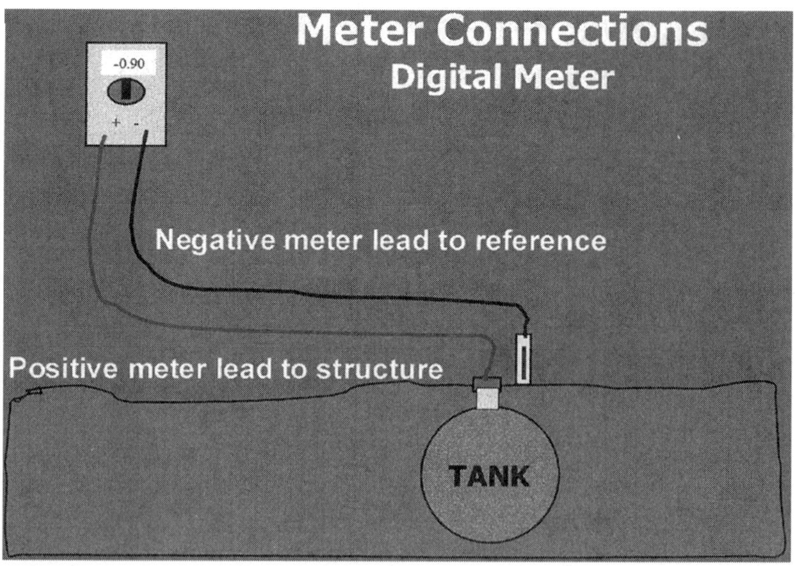

Figure 1 Meter connections: digital meter.

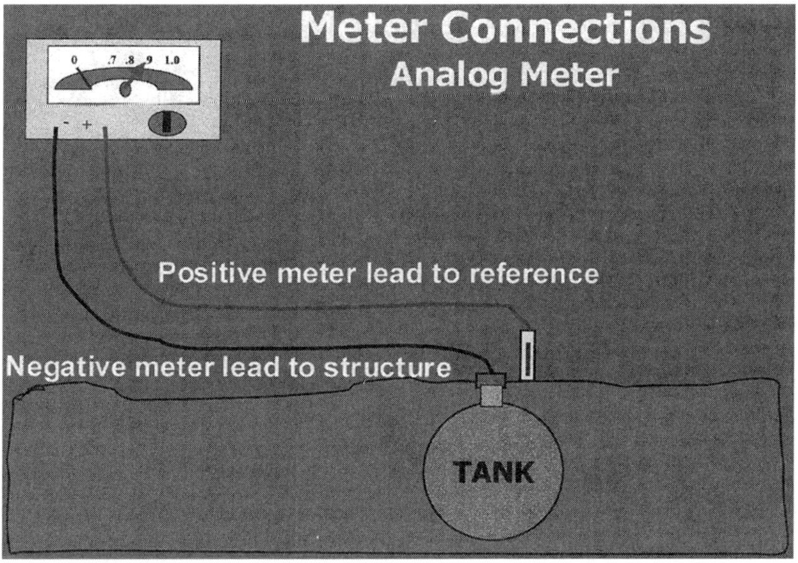

Figure 2 Meter connections: analog meter.

test and the negative meter lead (typically black) to the CSE. This will provide a numerical meter reading that has a negative polarity—for example, a readout of −.545 V.

The other approach, originally used with analog meters (Fig. 2), requires the positive meter lead to be attached to the CSE, and the negative meter lead to be attached to the structure under test. The reading obtained will be of a positive polarity, or .545 V. The reason the analog meter would appear to have the connections made in reverse is that the older analog multimeters did not have a large negative scale. As a result, connecting the meter to obtain a negative value (digital connection) would force the needle below the zero position on the scale. This action would fail to provide an accurate numerical value for the voltage measured. Both methods are acceptable; however, all values should be recorded using a negative polarity.

V. STRUCTURE CONNECTION

The structure connection may vary depending upon the existence of a test station and the technique used to measure the level of cathodic protection.

A. Test Stations

The typical structure connection is via a single test wire brought to grade in an accessible opening. The wire end at grade is secured to a riser pipe, but is not electrically connected to the pipe. The other wire end is electrically connected to the structure via a mechanical fastener or welded connection, and is protected from its own corrosion by coating and/or taping. This type of structure connection is used to perform an "on" potential measurement. The "on" potential is obtained without disconnection or interruption of the cathodic protection system.

Occasionally, the single test wire may be run to a terminal in an at-grade test station. Labeling or color coding should have identified the wire when the test station was installed. When this has not been done the wire must be identified through a process of elimination using potential measurements. If multiple wires are evident in an at-grade test station, the wires should be identified during testing and the information documented as a permanent record.

When measurements show that several wires have a similar electrical potential, consideration should be given to whether the structures connected to the wires are continuous with each other. In some instances potential measurements must be obtained directly from the various buried structures, such as piping, to identify the origin of any previously unidentified test wire. Additional information regarding continuity testing is provided in Section IX.

A test station is occasionally used to enable cathodic protection measurement by more than one method. This type of test station provides connection from the structure lead through a calibrated resistor (shunt) to the current source, typically a wire-connected sacrificial anode. This type of test station allows for disconnection of the anodes, which is required when using the 100-mV polarization method—a procedure not used with weld-on anodes.

B. Test Head

Another method of structure connection is by use of a test head, which is brought to grade. Among the most common of these is a buried reference electrode with connection points labeled on the test head. These points correspond to one or more cathodic protection test leads and a reference point that marks the connection to the buried electrode.

C. Without a Test Station

An alternate method to make connection to the tank when a single test wire cannot be located or identified, and no other means is available, is via the fill-pipe opening. This method calls for a temporary connection extended through the fill pipe until the tank bottom is contacted. This is most often accomplished by using a wooden gage stick with a wire secured along its length, and attached to a nail or screw inserted into the stick's probe end. The wire insulation is removed at each end to allow electrical connection to the nail or screw and a meter lead. Various other configurations of the same concept have been used, and range from a spool of wire with a weighted magnet (which is lowered into the tank to contact the tank bottom) to a plastic or PVC probe used similarly to the gauge-stick approach.

Each of the previously described test connection types will be discussed further during the test method section.

VI. REFERENCE ELECTRODE LOCATION

Prior to any electrical connection between the meter and the structure under test, the reference cell should be placed in its test location. The optimum location when testing an underground storage tank is above the tank, as close to the centerline as practicable and away from the anodes. This is typically accomplished via an at-grade opening in the concrete or asphalt.

The reference cell is positioned to embed the porous tip in the soil or backfill. It is most desirable for the tip to be firmly in contact with the backfill; therefore, twisting the cell during placement may be helpful. Before inserting the tip, an inspection of the area should be performed to assure that the cell is not placed in soil contaminated by hydrocarbons. The inspection may only require visual confirmation, although contaminated soil is most often detected by the smell of a petroleum product or by checking the soil for an oily texture. When the tip is placed in contaminated soil the pores may absorb some of the hydrocarbon or become blocked by pollutants. Because hydrocarbon products do not readily conduct electricity, the reading value obtained using a contaminated tip may not be accurate. If the tip is inadvertently placed in contaminated soil the tip should be replaced or thoroughly cleaned before further use.

Also, the cell tip should not be in contact with trash, debris, grass and leaves, or other metallic structures when a reading is obtained. If one or more of such items or conditions is present, another location should be selected for cell placement.

Use of a remote reference cell location is often preferred when one of the previously noted conditions exists, or because there are no usable openings above the tank. When using a remote reference cell, the location should be the nearest native soil that avoids the cause for cell movement, such as hydrocarbon contamination. It must be noted that placing the reference cell several feet to either side of the tank centerline does not constitute a remote location. Remote locations are typically more than 10 ft away from the tank excavation.

In no instance should the reference cell be placed so far away from the tank that other buried structures may interfere with the potential measurement. Here are two remote reference cell placement examples to avoid: on the other side of an adjacent street, or near a major utility such as sewer or buried electrical lines.

VII. "ON" POTENTIAL MEASUREMENT TECHNIQUE

Advantage: The "on" potential method is most often used for convenience since it is easily performed and requires less personnel and equipment than other methods.

Disadvantage: This method is limited in application and does not allow for meeting other criteria that are also recognized as evidence of adequate corrosion protection such as the 100 mV polarization criteria.

Once the reference cell has been placed in the soil, the structure and meter connections should be performed using the single-wire test lead or alternate tank connection, as previously described. The meter display will indicate the voltage reading measurement. That number must be recorded. When using the "on" potential measurement technique, the minimum criteria is −.85 V DC. That level and any reading more negative (e.g., −.9 V), indicates adequate protection from corrosion. A number more positive (e.g., −.7 V) would require further investigation.

VIII. 100 mV POLARIZATION TECHNIQUE

Advantage: May allow a tank with a potential less negative than −.85 V to be considered as adequately protected from corrosion.

Disadvantages: The technique only works with anodes that can be disconnected from the structure under test, and may require more than one person to perform the test properly. The 100 mV polarization technique is only performed when the cathodic protection current source can be interrupted or disconnected. When the current source is interrupted, the level of cathodic protection will drop immediately and suddenly. Because of this, strict attention must be given to the meter display at the instant the current is interrupted. The following procedure is used for the 100 mV polarization technique.

Situate the reference cell as previously described, and make the structure meter connection at the test station.

Disconnect the current source (or turn it off).

When interrupted, the meter reading on the display will immediately change to a smaller number and then pause. This pause may only be identified by the fact that a numerical value will actually be readable, momentarily, on the display.

The first value to be recorded is when the reading value pauses and is known as the "instant-off" reading.

The reading on the meter is then allowed to decay to its smallest value or until there is at least a 100 mV (.1 V) difference between the reading on the display and the reading obtained at the pause. This is the second value to be recorded. If a minimum 100 mV difference is observed between the first and second values, the tank is protected from corrosion.

If the reading obtained does not meet either of the criteria presented, the need for further testing is indicated. The additional testing is in the form of troubleshooting, specifically for continuity.

IX. CONTINUITY TEST

Many cathodic protection systems are designed to protect the tank only. To accomplish this it is necessary to electrically isolate the tank from other metallic structures or objects, such as piping. The electrical isolation defines the structure geometry to be protected by the cathodic protection. If the isolation should fail, be tampered with, or if an object is physically in contact with the tank (shorted), the cathodic protection performance will be impacted. Typically, the "shorted" structure is also electrically grounded and provides a surface area greater than the cathodic protection system was originally designed to protect. Breaking down the isolation results in a reduced electrical potential value for the structure originally intended for protection, such as an underground storage tank.

The continuity test—which can help determine whether a tank system is electrically isolated—is similar to the cathodic protection test, and can be used for any cathodic protection system. The reference cell is placed in a remote location and not moved until the test concludes. A tank potential reading is obtained and recorded. The positive meter lead connection is removed from the tank and touched to other metallic objects that could be connected to, or physically in contact with, the tank. This would include piping, conduit, pumping and gauging equipment, interstitial probes, etc. If a reading obtained from one of these objects is within 10 mV of the structure's reading, continuity is possible.

When continuity is detected and corrected shortly after installation there will likely be no detrimental impact upon the performance of the cathodic protection system. However, if allowed to go uncorrected, anode life can be severely diminished. When the continuity has been corrected an immediate improvement in the reading should be observed; however, it may take several days for the tank to meet the -.85 V criteria, if applicable.

It is important that test personnel have an understanding of the cathodic protection design when possible. Generally, the tank design information is available from the owner of the tank facility. This information will help the tester identify whether there should be isolation and/or continuity, as applicable.

Many field-installed cathodic protection systems designed to protect underground storage tanks rely upon continuity between the tank and piping to provide complete protection to the tank system. This system—an impressed current type—is usually installed on older, existing UST(s) and piping. An impressed current system can usually be identified by the presence of a cathodic protection rectifier that provides electrical power. For a system using a common cathodic protection source to protect both the tank(s) and piping, a lack of continuity between the tank and the product delivery line may mean the line is not fully protected from corrosion.

X. PRACTICAL DATA INTERPRETATION

Once the cathodic protection reading is obtained, it can be easily determined whether the system is meeting one of the minimum criteria. Beyond the reading and its relationship to the criteria, what does the number mean? This is a question that frequently arises when the value recorded does not meet one of the criteria presented. Although cathodic protection testing methods are not absolute science, the technology of cathodic protection has been proven effective in controlling corrosion. Monitoring is a widely accepted means of measuring cathodic protection performance. But the tank owners and testers should be aware that a low reading can sometimes disguise a cathodic protection system that is performing adequately. In many instances the reading is influenced by improper test techniques, changing soil conditions, or lack of familiarity with the tank system under test.

Many factors have to be considered when attempting to interpret a reading value. First and foremost, a thorough knowledge of the structure, its cathodic protection system, and the performance features, is essential. A reading obtained from a bare (uncoated or poorly coated) structure should not have the same interpretation as the same value obtained from a well-coated, cathodically protected tank.

The reading value obtained when a metal is tested with a half cell (CSE) is actually the difference in electrochemical potential (voltage) between the reference device and the metallic object under test. The more similar the metallic object is to the copper cell, the smaller the electrochemical difference, thus a smaller numerical value. For example, a potential value obtained from a copper rod with reference to a copper sulfate electrode would typically be a very small number. The value could be as low as .15 V. The small value is because of the electrochemical similarity (copper compared to copper). However, it is rare that the value is actually 0.0 V due to the difference in purity between the copper rod in the ground and the copper rod in the half-cell sulfate solution.

Test personnel can expect to encounter certain voltage levels on a routine basis when testing cathodic protection on tanks. The ability to identify a structure with some degree of certainty can be valuable. Often, the value itself can be used for identification. Clean mild steel will typically yield a potential value between

−.65 V and −.6 V, although it may be as low as −.5 V in excessively dry soils. Zinc itself will typically have a maximum potential value of −1.1 V. Magnesium can read as high as −1.8 V. However, zinc and magnesium values may be lower depending on their age, purity, and numerous other variables.

XI. TAKING THE VALUE FOR GRANTED

It is often easy to record a value and assume you know what it means. However, there can often be deceptive circumstances surrounding the reading value. Nonmetallic fiberglass-reinforced plastic (FRP) tanks can provide values as low as −.2 V. Similarly, steel tanks with a lining can provide values in the same range if an internal connection is used in lieu of a single wire test lead attached to the tank exterior. If the tester is not familiar with the vessel—e.g., not knowing whether the tank is cathodically protected, composite or employing some other corrosion protection technology—the lined steel tank's reading value could be inaccurately interpreted as (1) active corrosion on a bare steel tank, (2) a steel tank connected to copper, or (3) possibly a fiberglass tank.

All too often, a reading evaluation leads to an assumption that the anodes on a cathodically protected tank have been consumed, although the tank is properly installed and isolated. This assumption is based upon a hasty interpretation of a low value such as −.4 V, which in many instances could mean the active presence of corrosion. Real-world testing experience has found that the value of −.4 V can also be related to extremely dry, nonconductive soil conditions in which the soil's resistance is so great that electrochemical current flow is inhibited. This can result in readings that seemingly convey ongoing corrosion, which would lead to a false conclusion on the need for corrective action. It is commonly found that a value as low as −.4 V can be caused by anodes that did not have the plastic outer wrap removed at installation. In this case, the corrective action would be to remove the plastic. However, this is not practical in most instances because of tank burial depths and the need for partial excavation. If that is the case, supplemental anodes can be added to the system.

As noted earlier, high-resistance soil can have a large effect on the reading and subsequent evaluation of the system performance. To properly evaluate an active cathodic protection system it must be tested while it is in an active performance mode, which would be defined as moist or wet soil. Such soil conditions should apply to both the current source (anode) and the reference point (CSE). If testing is performed during dry conditions, an attempt should be made to introduce water into the excavation.

There is no pat answer to the questions of how much water to add and how long to wait before testing the system after introducing moisture. Some locations require a lot of water because of rapid drainage, or because the anodes were never wetted at installation, not just because it is dry. The amount of time between introduction of water and a reaction from the anodes depends on several factors also, such as the tank size, backfill type, coating condition, etc. The CSE placement location should also be moist or wet. Another situation to avoid is testing in extremely frosted or frozen soil, which will significantly resist current flow.

Knowing the typical potentials for metals—as represented in a table of the galvanic series—can aid in determining what a reading value may mean (Fig. 3). For example, when mild steel that is under cathodic protection is shorted to copper the level of protection can easily be diminished by 600 mV. This can result in cathodic protection monitoring values near −.4 V or less. A cathodic protection reading measures the influence of an anode on a steel structure. By the same token, an electrical potential reading taken when a dissimilar metal is electrically connected to the cathodic protection system may help identify the culprit.

Here is another potential pitfall. A test lead that is improperly connected to the tank will not provide an adequate electrical circuit for low voltage measurements. A bad connection could result in a low reading and the misevaluation of a properly performing system. As a helpful hint, test leads with buried connections should always generate suspicion. An additional reading should be obtained using an alternate connection method whenever possible, such as those discussed in Section V.C. This will help identify whether the low reading is actually due to a poor electrical connection to the structure or whether the system is providing substan-

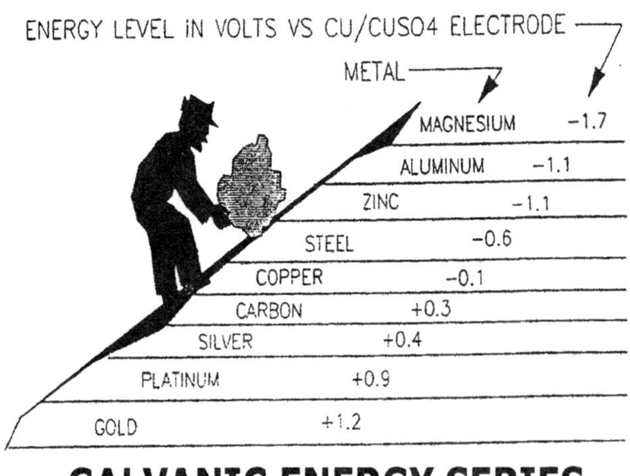

Figure 3 Nature has endowed each metallic substance with a certain natural energy level or potential.

dard performance. When the reading is similar to that of copper it is likely that the test wire is not connected to the structure.

XII. INTERFERENCE

Readings are often taken above the tank in an opening that is provided for access to piping, gauging or pumping equipment. When this reference location is used several cautions must be remembered. First, always be sure that the soil where the reference cell is located is the same soil surrounding the tank. If a containment sump holding soil or rock is used for the reference cell placement, the potential will not be the actual reading of the tank. If a reading obtained from such a location is unusually low, or fluctuates erratically, another reference location must be selected to verify the reading.

Sometimes the reference location is between—or physically near—a rusted pipe or pump and the at-grade-opening ring. Rusted metallic objects have an area of influence in the soil that is created by their own electrochemical corrosion process. If the reference is located within the area of influence, the active corrosion current can provide a false reading of the cathodic protection system under test. If this scenario exists or is suspected, an alternate reference location should be selected.

XIII. OTHER VARIABLES

Other factors can influence the reliability of the test results. The short list below will attempt to address a few.

The tester's hands should never touch the meter connection clips or other wire clips during the test. The inductance provided by a human body will alter the reading accuracy. Also, wire connectors should not be allowed to rest on moist earth or on wet surfaces.

A reference electrode that is frozen or contains a slushy solution should not be used for a cathodic protection test. The low temperature will affect the rate of electron flow within the cell.

The window on a reference electrode should not be exposed to direct sunlight for long periods of time. The copper-sulfate solution is photosensitive and will be affected by prolonged exposure to the light.

Obviously, all of the possible scenarios that could affect the accuracy, or meaning, of the measured voltage on a cathodic protection system cannot be discussed here. The previous paragraphs discussing the reading interpretation have been provided to illustrate the point that the number by itself can often be misleading.

When a potential that does not meet one of the criteria is measured, troubleshooting should be performed to help identify the cause. Often, every aspect of the installation and test procedure requires review.

Occasionally, retesting the system is necessary to verify previous results or to perform troubleshooting. During the retest, every effort should be made to remedy situations or causes that are known to have an impact on the cathodic protection reading accuracy. For example, if a dry soil condition is suspected as the cause of a low reading, re-test when soil conditions are more favorable (such as after normal seasonal rains). The tester should be familiar with the influences that can result in inaccurate voltage readings. Once aware of these influences, the tester can use additional testing to verify previous results, or identify, to a certain degree, the cause of the reading.

16
How to Specify UST System Equipment

Alex Ralston
Petcon, Inc., Jackson, Mississippi

The equipment mentioned in this section is based on federal underground storage tank (UST) regulations, 40 CFR 280. These are minimum standards, which means a state has the option to make them more restrictive. Always check with state and local UST regulators overseeing your site before specifying any type of tank-system equipment. A conversation with environmental and fire-code regulators can save time and money on a project because sometimes site-specific conditions do not allow the use of a particular method or piece of equipment.

I. PIPING: PRESSURIZED, SUCTION, VENT, OR VAPOR RECOVERY

Piping is one of the most critical pieces of the overall tank system. Studies have shown that piping often has been the true source of contamination in situations referred to as "tank" leaks.

A. Pressurized Piping

1. Single Wall

Single-wall piping is used at locations where secondary containment is not required or in areas not sensitive to environmental contamination. It is usually less expensive than double wall piping.

A single-wall piping layout may be used on suction systems with only one check valve in the system directly under the pump—with the product line sloping toward the tank and no other check valves in the system. This system does not pose as large of an environmental threat as a pressurized system because a piping leak will cause the product to flow back into the tank instead of into the surrounding soil. Suction systems with a check valve located at the tank are required to have some other method of leak detection.

Single-wall pipe is often used on vent lines since they pose less of an environmental threat than product lines and do not normally contain liquid product. Most regulators allow vent lines and vapor recovery lines to be constructed of single-wall piping.

2. Double Wall

This type of piping is used at environmentally sensitive sites or sites where secondary containment is required. These are typically areas where the local drinking water source is very near the site.

Some types of piping—such as flexible direct-bury pipe—is only sold in double-wall construction. The second wall provides a margin of prevention against contaminating the environment. Pipe monitoring must occur at least once every 30 days. Records must be kept of monitoring results. Either the interstice—the space between the pipe's two walls—or the containment sumps to which the pipe is connected must be monitored.

B. Suction, Vent, or Vapor Recovery Piping

This pipe must be of rigid construction so that it does not have any swags or low places. Flexible pipe is not suited to this type of system unless it has been specifically designed for this purpose. *Note*: Some flexible pipe manufacturers also make pipe that may be used for suction systems and vent lines.

C. Metallic or Nonmetallic

1. Metallic

Because of its resistance to fire, metallic pipe, such as galvanized steel, typically is recommended on aboveground vent risers. However, the underground use of galvanized piping is prohibited for UST systems (unless it is used in conjunction with corrosion protection methods that meet regulators' approval) because the zinc coating by itself is inadequate to provide reliable, long-term corrosion protection. Metallic piping, including flex connectors, may be used underground—but must be coated with a dielectric material, compatible with the product stored and cathodically protected. Field-installed cathodic protection systems must be designed by a "corrosion expert," as defined by 40 CFR 280.

Very little metallic piping is installed today for petroleum underground product lines because the nonmetallic piping is lighter, easier and faster to install than metallic piping. For chemical compounds other than petroleum, check with the piping manufacturer to ensure that the fluids to be stored and transferred are compatible with the piping material.

2. Nonmetallic Piping

Nonmetallic piping is widely used for all underground petroleum piping applications such as product lines, vent lines, and vapor recovery lines. In most cases, it takes less time to install than metallic piping and is easier to handle.

Nonmetallic piping is noncorrosive so no additional coating or cathodic protection is required. Flexible piping has few joints, which are made inside a containment sump—decreasing the likelihood of a leak entering the ground. Check the Underwriters Laboratories listing for compatibility with alcohols and chemical compounds to be stored. Many nonmetallic piping manufacturers can fabricate piping compatible with 100 percent methanol, for example.

a. Fiberglass

Advantages: This material can accommodate sharp 90° bends where flexible piping requires more space with long sweeping bends. Its rigid construction makes it desirable for use on vent lines, suction lines, and vapor recovery lines.

Disadvantages: Longer installation time. More steps may be required. Assembly may be slowed when working in extremely cold temperatures.

b. Flexible Piping.

Advantages: Several factors have increased the popularity of flexible piping systems: fast installation, the elimination of fittings or enclosure of fittings in a containment sump, and the overall design. One-piece coaxial construction allows direct burial of double-wall pipe.

Disadvantages: However, flexible piping has limitations. Its flexible construction makes some types of this piping unsuitable for vent or vapor recovery lines. Long sweeping bends require more room at an installation site than rigid piping. The cost of materials can be more than rigid piping.

II. LEAK DETECTION

A. Tanks or Piping

All regulated tanks must have at least one approved method of leak detection installed. Refer to 40 CFR 280 for all approved methods. This chapter will outline the most widely used equipment. Other methods of leak detection are allowed. However, some can be used for only a limited number of years. Some do not require any specific equipment (e.g., statistical inventory reconciliation [SIR], inventory control, tank tightness testing, and manual tank gaging). *Important*: Leak detection equipment must have documentation of third-party certification signifying that it meets EPA's standards for detecting a leak. Any equipment or process without third-party certification will not comply with the UST leak-detection regulations.

1. Tanks

a. Automatic Tank Gaging. Automatic tank gaging systems vary from the very basic to highly complex, featuring options such as high and low-level alarms, interstitial monitoring, automatic leak testing, modems for remote data transmission, and many others.

For automatic tank gaging to be accepted as a leak detection method, the most important facet of meeting UST requirements is the monthly leak test. This test can be performed manually by shutting down the system for about four to six hours. During that span, no product can enter or leave the tank as the test progresses. At the end of the test the equipment will provide the operator with results. This test may also be performed automatically without having to shut down the system completely.

Be sure that the manufacturer has third-party certification on all types of leak tests.

b. Interstitial Monitoring: Manual or Electronic. Either of these methods requires monitoring at least once every 30 days and appropriate record keeping.

Manual Interstitial Monitoring. Some double-wall tanks have a liquid-filled interstitial space where the level of the liquid can be monitored for leaks. Other tanks provide a dry interstitial space that can be monitored monthly for the presence of any liquid.

Automatic Interstitial Monitoring. This method uses sensors to either indicate the presence of liquids or vapors in the interstice or record any variations in the liquid level. Even though the monitoring is done automatically, the monitoring must be done at least every 30 days and test results properly recorded.

Groundwater and Vapor Monitoring. These are special wells installed within the tank excavation. An assessment must be conducted to determine if the site is suitable for groundwater or vapor monitoring. Each state has its own specific regulations on the use of groundwater and vapor-monitoring wells, which are not universally accepted. Groundwater monitoring can only be used if groundwater is within 20 ft of the surface and the backfill surrounding the tanks and the wells is porous enough to allow the free flow of liquid.

A vapor-monitoring system can only be used to keep tabs on volatile products such as gasoline, some organic solvents, and, in some cases, diesel fuel. It cannot be used to monitor low-volatility products such as motor oil.

A special instrument that gives readings in parts per million (PPM) must be used to conduct vapor monitoring. The instrument can be either portable or part of a fixed, electronically monitored system.

2. Piping: Suction, Pressurized, Vents, or Vapor Recovery

a. Suction Piping. If a check valve is installed directly below the pump, with no other check valves in the system and the piping sloped toward the tank, no additional leak detection is required. Also, this design requires no monthly monitoring on the suction piping only. If such a design is not used, then one of the following leak-detection options must be employed on the suction system:

A tightness test every three years
Interstitial monitoring
Groundwater monitoring
Vapor monitoring
Statistical inventory reconciliation (SIR)

(*Note*: Only some methods of SIR are accepted by states. Check with the state UST regulatory division.)

b. Pressurized Piping. Because pressurized piping has the potential to spill massive amounts of product, two methods of leak detection are required (Table 1).

c. Vents and Vapor Recovery. Since these pipes do not routinely contain product, leak detection is not required. It is extremely rare to find tank systems employing leak detection on these lines.

III. SPILL PREVENTION

Each regulated UST is required to be equipped with a spill containment manhole, which at a minimum will hold 5 gal of product. They come in a variety of sizes, and include options such as drain valves and small hand pumps that remove any product collected in the manhole.

Table 1 Pressurized Piping Leak Detection Requirements (Employ Two Methods; One from Each Column)

A device that restricts the Øow of product	An annual tightness test on the pressurized piping
A device that triggers an alarm	Monthly vapor monitoring
A device that shuts off the Øow of product	Monthly groundwater monitoring
	Monthly interstitial monitoring

Source: U.S. Environmental Protection Agency.

IV. OVERFILL PREVENTION

Each regulated UST must be equipped with an overfill prevention device that will allow the tank to be filled no higher than 90% or 95% of capacity, depending on whether the model shuts off the flow of product into the tank. EPA also allows an alternative method that permits larger tanks to be filled to a greater capacity if certain conditions are met.

A. Ball Float Vent Valve

On the plus side, this valve can be used for two-point Stage I vapor recovery—and it's less expensive than overfill prevention valves that fit into drop tubes. However, the ball float vent valve is not recommended in the following situations:

Suction pumps
Where deliveries are made into the tank by pumping rather than gravity drops
Where a coaxial Stage I vapor recovery system is used
Remote fills
Emergency generator tanks

Another potential concern is that the ball float valve does not permit a tank manager to utilize storage capacity as completely as other systems. The ball float valve must be set at 90 percent of tank capacity, rather than the 95 percent level allowed on positive shutoff systems. In addition, if a ball float valve is installed in fittings on top of the tank, it is not suitable for retrofitting without incurring the many costs of breaking pavement.

B. Overfill Prevention Valves with a Positive Shutoff

Not all specifiers prefer the positive shutoff feature for overfill prevention valves. Price has something to do with their feelings. The product is about triple the cost of a float vent valve. In addition, it is not recommended for use where deliveries are made by pumping into the tank rather than by gravity drops.

On the upside, the valve enables the tank to be filled reliably to a maximum of 95 percent of tank capacity. This valve also is easy to install in the tank fill pipe. Additionally, the overfill prevention valve can be used when the ball float is not recommended.

C. High-Level Alarms

Tank managers have found that an alarm signaling a potential overfill condition has some drawbacks:

An inability to restrict, or shut off, the flow of product into the tank

The need to locate the alarm outdoors, near the tank, where it would be heard by the transport driver (and, unfortunately, by customers)
Electrical wiring installation, not required by mechanical methods
Must be set at 90 percent of tank capacity, except where an approved alternative method is used
Frequent triggering leads to complacency and ignoring the alarm, much like car alarms

In the right situation, a high-level alarm enjoys some advantages:

Can be an option on an automatic tank gaging system
Ability to work with either suction or pressurized systems
Can be used where deliveries are made into the tank by pumping

V. CONTAINMENT SUMPS: TANKS AND DISPENSERS

A specifier has several materials from which to choose. Sumps are available in a variety of plastic configurations, fiberglass, and steel. The key to a successful dispenser-sump installation is ensuring that the sump is liquid-tight to prevent migration of any hazardous materials into the environment as well as any water entering the sump. Similarly, the primary consideration for tank sumps is liquid-tight installation if the structure will be used as a termination point for secondary containment piping.

Tank and dispenser sumps fabricated of plastic or fiberglass require very careful backfilling because the backfill becomes critical to providing structural strength for the sump walls. For best results, follow the manufacturer's installation instructions.

17
AST Design

Lorri Grainawi
Steel Tank Institute, Lake Zurich, Illinois

I. A BETTER WAY TO COMPLY

The proliferation of aboveground storage tank (AST) designs during the 1990s truly reflects marketplace demand from disgruntled owners of underground storage tanks (USTs) to creative facility managers trying to lower costs while ensuring that petroleum and chemical storage needs continue to be met. The decision on how to store flammable, combustible, or hazardous materials in capacities of 50,000 gal or less has never been more complicated. After all, for decades, there was typically one answer—underground storage.

Today, an aboveground tank often can take the place of an underground storage system. The AST design can be single wall, single wall in a dike with or without a rain shield, double wall, double wall with insulation between the two walls of steel, single-wall steel with a concrete secondary containment chamber, and on and on. And all the designs have some logical basis. Some, of course, are more logical than others.

But the beauty and the logic of each design are in the eye of the beholder—or in the mind of the specifier. After all, the true name of the game in AST technology is compliance—most often with fire and building codes, or standards developed by nationally recognized testing laboratories.

The job of specifying requires the ability, among other things, to satisfy the demands of tank owners and regulators who individually could be concerned about:

The AST system's impact on navigable waterways
Prevention of releases to avoid fires
Tank ventilation

Tank system proximity to buildings and property lines
Emissions of volatile organic compounds
Ballistic-impact resistance
Barrier provisions to repel a motorized vehicle that skids on an icy driveway

To manage the variety of regulatory issues, Underwriters Laboratories (UL) and Southwest Research Institute (SwRI) during the 1990s developed testing procedures that matched the emerging demands from fire-code revisions and the needs of environmental regulators. New terms for ASTs surfaced, such as "fire resistant," "protected," and "multihazard."

Here is a look at some of the key design features for aboveground storage tanks.

II. UL 142

This UL standard (Steel Aboveground Tanks for Flammable and Combustible Liquids) covers the manufacture of the vast majority of steel shop-fabricated ASTs in America. Whether the tank will be double-wall design, single wall unitized with a dike, or some other innovative use of steel, the basic, primary storage tank will be governed by UL 142. (For a look at AST standards' requirements in Canada, see Chapter 19).

The UL 142 standard specifies mandates for a wide variety of fabrication issues (e.g., weld profiles, fittings, bulkheads, steel gage thickness, manways, leakage tests, etc.) For more information, see Chapter 18.

A. Cylindrical

The cylindrical design is the most common among UL 142 tanks. It can be adapted for either horizontal or vertical tank applications. Cylindrical ASTs can be strictly steel or combined with other materials for situations that require protected tanks. When specifying tanks of any size, the cylindrical design is most often selected.

B. Rectangular

Rectangular ASTs are most often used in conjunction with protected tanks or workbench lube oil applications. They are typically smaller capacity (5000 gal or less) because a cylindrical tank design is considerably more economical in larger sizes. To receive UL approval, rectangular tanks must undergo specific performance testing, such as a top-load test of 1000 lb.

C. Single Wall

Prior to 1990, this was virtually the only way small-capacity ASTs were fabricated. However, marketplace demand, driven by environmental regulations and changes to the fire and building codes, has significantly diminished the importance of standalone, single-wall designs. Single-wall tanks still outnumber ASTs built with secondary containment because of the great number of tanks installed at farms, construction sites, and remote locations—or in basements or small, outdoor heating-oil applications. However, stricter rules in fire codes and environmental regulations have increased demand for AST secondary containment systems.

D. Secondary Containment

The trend toward secondary containment has grown dramatically in the past 10 years, which also has spawned a flurry of unique designs. Because of site-specific conditions, a specifier cannot assume that any secondary containment AST will unconditionally satisfy a regulator or inspector. Always check with the authority having jurisdiction (AHJ) before investing in the purchase and installation of an aboveground storage system.

Another key factor to consider is the overall weight of an AST system with secondary containment. Depending upon the design, a 10,000-gal AST can weigh anywhere from 15,000 lb to more than 90,000 lb. Heavyweight designs can pose both logistical and cost challenges for transporting the tank.

1. Single-Wall Diked

A single-wall AST placed in a dike is an elementary form of secondary containment. The primary purpose of the dike pertains to satisfying fire code requirements—either containing a fire near a tank or preventing flammable liquids from creating a hazard by flowing toward other buildings or properties.

For many years, an earthen dike was the commonly accepted method of surrounding a single-wall aboveground storage tank to provide secondary containment. As concerns about soil contamination increased, the need for more impermeable containment solutions escalated. For example, the U.S. Environmental Protection Agency (EPA) proposed a 1991 revision to the Spill Prevention Control and Countermeasures (SPCC) program that recommended the use of secondary containment barriers capable of remaining impermeable for 72 hr. Granular soil, such as sand or crushed stone, cannot meet such a specification.

The single-wall, open-top diked design enjoys some popularity, especially for capacities under 2000 gal.

Dikes have several benefits for AST owners. A dike can contain spills from pipes and valves. If the AST would somehow catch on fire, a dike can serve as a coolant tub—collecting water that moderates the tank's temperature as firefighters battle the blaze. And a dike can be a meaningful collision barrier against damage to the primary tank.

However, an uncovered dike—whether it's made of steel, soil, or another material—also serves as a receptacle for precipitation. If the tank is overfilled when water has already accumulated in the dike, the AST manager will have to deal with removal of a hazardous substance—and the various regulations that govern the handling and disposal of such liquids.

Many manufacturers have developed rain shields and other, similar designs to prevent the accumulation of moisture in the diked area. Some manufacturers have designed their systems so that overfills are directed into the diked area, rather than over the rain shield and onto the pavement or nearby soil. Rain shields also are available in two primary designs—hinged top or welded solid. The latter must be equipped with an emergency vent. Fire codes require all enclosed areas that may contain flammable liquids to be capable of venting vapors during a fire. Some rain shield diked designs incorporate staircases and platforms to make tank filling easier.

Many dikes for AST secondary containment are designed to hold 110 percent of the primary single-wall tank's capacity (although a few states require 125 percent). Such a design deals with the potential for capturing precipitation in the diked area, particularly if a means to prevent overfills is not adequately addressed through shutoff devices, alarms, and gages. This means a larger area for installation is required, which can increase costs. Liquids from the dike must be removed to maintain the secondary containment capacity.

2. Double Wall

The double-wall AST design was developed around 1990 as a close relative to the dual-wall UST that was first fabricated in the United States during the mid-1980s. In most cases, it's two walls of steel in an intimate-wrap approach.

The early years of double-wall AST usage were marked by some skepticism that the system would truly contain all releases. However, as the industry improved methods for combining the double-wall tank design with enhanced technology for containing and preventing overfills, many authorities having jurisdiction and tank owners grow more comfortable with the approach.

As rain shields became more common (to prevent the infiltration of water) and thus an inherent part of the diked AST system design, the practicality of double wall ASTs with overfill prevention devices became more apparent.

There are several benefits to double-wall tanks over diked AST Systems:

No place for rainwater to collect

Flammable and combustible liquids are contained without exposure to ignition sources, thus reducing the potential of fire

Secondary containment is testable to prove integrity and assure the containment's performance

A smaller footprint, which requires less real estate for installation and operation

Greater accessibility to the tank's top

E. Triple Wall

Tertiary containment systems have been manufactured in rare cases. Most specifiers shy away from triple-wall systems because of the added cost. But in some areas where the potential for seismic activity exists—or the protection of aquifers is considered the top priority regardless of cost—triple-containment systems have been placed in operation.

III. FIRE-RESISTANT AND PROTECTED TANKS

As demand has increased for the use of ASTs in the fueling of vehicles, the major code development organizations have created one of the newest categories of aboveground storage—tanks that can withstand the intense heat generated in a pool fire that lasts for two hours. Manufacturers of two-hour, fire-rated tanks have demonstrated to nationally recognized testing laboratories that their products can maintain structural integrity during an extended fire. Additional requirements may include ballistics and simulated vehicle impact tests to demonstrate the AST's rugged design.

A. Fire Resistant

The National Fire Protection Association's NFPA 30 (Flammable and Combustible Liquids Code) coined the term "fire-resistant tank" in the early 1990s. The new category of aboveground storage tank was created in response to states that were enacting new laws that allowed ASTs in lieu of USTs for fueling without requiring proper safeguards. Fire code officials, alarmed at the public safety implications, determined that new mandates for AST fueling systems were needed. So language was promulgated to assure safe design and installation. NFPA desired a tank that would not fail or collapse during a strenuous two-hour test featuring exposure to high temperatures. As the fire codes adapted to the new realities of growing demand for ASTs, the nationally recognized testing laboratories responded.

UL Subject 2080 (Outline of Investigation for Fire Resistant Aboveground Tanks for Flammable and Combustible Liquids) requires a two-hour fire test in which a maximum single-point temperature is allowed to reach 1000°F and the average temperature throughout the internal tank is 800°F.

The Southwest Research Institute (SwRI) Test Procedure 97-04 calls for an aboveground tank to meet fire-resistant requirements of NFPA 30A (Automotive and Marine Service Station Code) without requiring temperature limitations for the primary tank during a two-hour fire test [1].

B. Protected

The "protected tank" concept originated within the Uniform Fire Code, which did not allow ASTs to be used for motor vehicle fueling, except for one situation. The code allowed small fueling tanks to be placed within a special concrete enclosure in buildings. A rationale emerged: If this was allowed, why couldn't adequate safeguards be created for outdoor ASTs, too? The Uniform Fire Code developed language striving to emulate a UST installation in which soil fully insulates the tank from fires. As they did with fire-resistant tanks, UL and SwRI came up with new standards that addressed the fire code changes.

UL 2085 (Protected Aboveground Tanks for Flammable and Combustible Liquids) requires a two-hour fire test at 2000°F during which an AST's maximum single-point temperature is allowed to reach 400°F and the average temperature rise throughout the internal tank can be no greater than 260°F. This standard allows the tank to be investigated to determine acceptability for use after damage from fire exposure, collision, or misuse.

The SwRI Standard 93-01 (Testing Requirements for Protected Aboveground Flammable Liquid Fuel Storage Tanks) parallels UL 2085 in mandates for a full-scale fire test, a hose stream test, and leakage testing. Each of the standards differs on requirements for vehicle impact, projectile, interstitial communication, fire testing the interstitial insulation, and several other tests.

C. Multihazard

One of the newest AST fabrication categories is a multihazard design (SwRI 95-03). This test procedure enables the reevaluation of a tank to determine if it can remain in service after experiencing damage from a fire, a vehicle collision, or some other destructive source. SwRI 95-03 requires a fully assembled tank to be exposed to the same tests required for a protected tank.

Afterwards, a manufacturer must refurbish the tank to like-new condition and expose it to another two-hour fire test at 2000°F. The AST will meet SwRI 95-03 requirements if the tank can again pass the test criteria for a protected tank.

IV. VAULTED TANKS

Another new standard is UL 2245 (Below-Grade Vaults for Flammable Liquid Storage Tanks), which was published in February 1999. This type of storage system is composed of a UL 142 tank enclosed in a concrete vault. The standard addresses leak detection monitoring, normal venting, emergency venting, and tank testing.

V. BINATIONAL STANDARD

UL and Underwriters' Laboratories of Canada (ULC) have been working to develop a new binational standard for protected tanks. To accomplish this, an industry advisory committee was assembled in 1996. An initial review noted the many similarities between manufacturing requirements in each country. Among the requirements included for consideration in the first draft are:

Ballistics and impact resistance for the primary tank, but not the secondary tank
Two-hour fire test temperature limitations for a protected tank, (i.e., no more than a 260°F average temperature rise on the primary tank)
Tank construction complying with an approved standard such as UL 142, or ULC-S601
Venting performance determined by UL 142, as this is the most stringent standard [2].

VI. UL 2244

This is a listing for an aboveground shop-fabricated storage tank system, which is a tank complete with all equipment necessary to make it both functional and compliant with national fire codes. All systems are based on secondary containment. According to UL, this standard was developed after requests by AHJs for UL to evaluate complete tank systems. The UL 2244 listing simplifies the approval process for the AHJ.

At the time a UL 2244 system is installed, a fire code or building code inspector can simply check the Code Compliance Verification List developed by UL. This list must be shipped with each UL 2244 tank system.

To qualify for use on a 2244 system, all equipment must be evaluated by UL. Some of the equipment, such as the tank, pump, and emergency vents, for example, must be UL listed (Fig. 1). Other accessories, such as a normally open vent or block valve, do not require UL listings.

A UL 2244 listing is available to any organization or individual that agrees to periodic UL visits at the factory location where the tank system is assembled before shipment to the installation site [3].

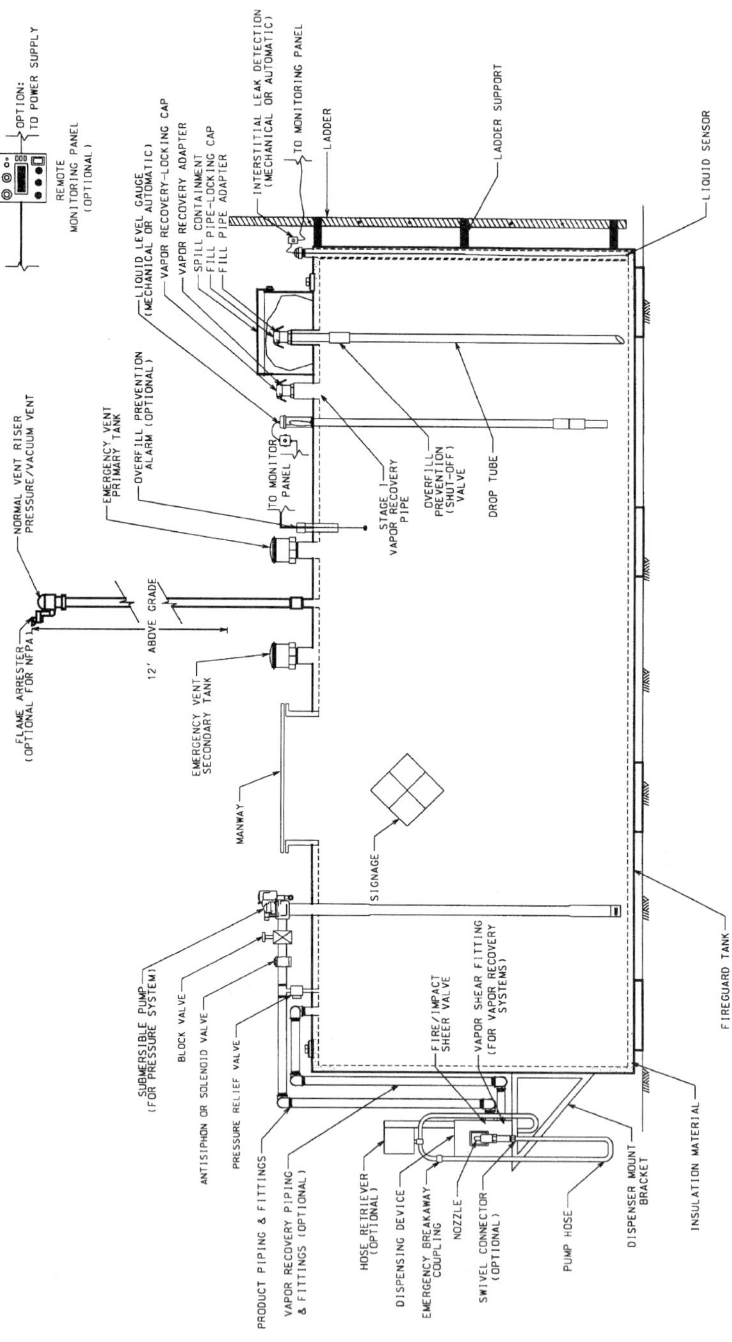

Figure 1 Motor vehicle fuel-dispensing system with side-mounted dispenser with vapor recovery. Courtesy of Steel Tank Institute.

VII. UFC APPENDIX II K

The Uniform Fire Code (UFC) adopted new provisions for aboveground storage tanks on August 17, 1998. The provisions are in the form of an appendix, which means each jurisdiction that abides by the UFC will have the option of adopting the new requirements. These provisions allow a UL 142 storage tank to be used for dispensing fuels at nonpublic service stations. The appendix provides guidance for the AHJ when approving an AST, such as requiring, among other things:

Spill and overfill protection
Warning signs
Minimum setback requirements from the tank to the dispenser, important buildings, and property lines

The idea behind all of these requirements is to protect both life and property, particularly other people's property.

REFERENCES

1. G Finley. Certification and testing of aboveground storage tanks. Proceedings of Aboveground Storage Tank Symposium. Atlanta, GA: Atlanta Fire Department, 1998, p 10.
2. Steel Tank Institute. UL Issues 1st proposed edition of new UL 2244 system listing. Tank Talk 13(3): p. 1, 1998.
3. Steel Tank Institute. UL 2244 system listing: new listing for complete fuel system package. Tank Talk 12(5): p 6–7, 1997.

18

Development of UL Standards for Aboveground Steel Tank Safety

Shari Hunter
Underwriters Laboratories Inc., Santa Clara, California

John J. Hawley
Underwriters Laboratories Inc., Northbrook, Illinois

I. INTRODUCTION

As was noted in the Underwriters Laboratories (UL) chapter in this book's underground storage system section, UL evaluates more than 80,000 products in more than 17,000 categories annually, including many items for the petroleum equipment industry. The rapidly expanding arena of small-capacity aboveground storage tanks (ASTs) is no exception.

UL Standards for Safety for aboveground steel tanks intended for the storage of flammable and combustible liquids include:

UL 142 (Steel Aboveground Tanks for Flammable and Combustible Liquids)
Subject 2080 (Outline of Investigation for Fire Resistant Aboveground Tanks for Flammable and Combustible Liquids)
UL 2085 (Protected Aboveground Tanks for Flammable and Combustible Liquids)
UL 2244 (Aboveground Tank Systems for Flammable Liquids)
UL 80 (Steel Inside Tanks for Oil Burner Fuel)
UL 433 (Steel Auxiliary Tanks for Oil Burner Fuel)

UL standards are developed based on the needs of state and federal regulatory authorities. During the development of safety standards, UL solicits advisory information from regulators, manufacturers, trade organizations, and others interested in the specific standard. Additional information on the various processes by which standards are developed is in Chapter 9.

Here is a summary of the key distinctions between UL standards that govern the manufacture of safe ASTs.

II. UL 142

This standard covers steel aboveground storage tanks. In 1922, the requirements for aboveground tanks were transferred from the National Board of Fire Underwriters (NBFU) regulations (Installation of Containers for Hazardous Liquids) and published in the first edition of UL 142. The safety standard addressed issues relating to leakage, venting (normal and emergency), and the ability of the tank to withstand the development of internal pressures encountered during leak testing. UL 142 covers tanks designed to operate at atmospheric pressure. Aboveground tanks include:

Horizontal cylindrical
Vertical cylindrical
Rectangular types

These are primary containment tanks (single-wall steel). They may optionally provide secondary containment by constructing the primary tank within a secondary steel tank or by constructing the primary steel tank within a steel dike intended to contain product from spill, leakage, or rupture. Additionally, the tanks may be provided with integral supports. In some cases, a UL 142 tank may be part of another UL-listed AST design, e.g., a cylindrical steel tank fabricated for use within a concrete containment chamber. In this particular situation, UL 2085 would also apply [1].

The requirements for horizontal and vertical cylindrical tanks are construction based, such as minimum steel thickness, maximum diameter, maximum length-to-diameter ratio for horizontal tanks, maximum height of a vertical tank, types of joints, minimum size of vents, etc.

The requirements for rectangular tanks are performance based. The standard specifies a minimum steel thickness and requires that the tanks be subjected to performance tests to demonstrate the strength of the assembly and the welded joints. The tests examine leakage, hydrostatic strength, and the top load. During the leakage test, the tank is pressurized and checked for weak spots (from which air or fluid could escape) and permanent deformation. The hydrostatic strength test is conducted by pressurizing the tank to five times the "marked" leakage test pressure.

The tank shall not rupture or leak during this test. Flat-top tanks are subjected to the top-load test. A 1000-lb load is placed on the weakest area of the tank top, after which the tank is checked for leakage.

Diked tanks are subjected to a buoyancy test. For the buoyancy test, the dike area is filled with water to its maximum capacity while the tank remains empty. The tank shall not float away from the dike. The dike is then emptied and checked for evidence of structural damage. Immediately following the bouyancy test, the tank is subjected to the hydrostatic load test, after which the dike area is again filled with water and checked for deflection. There shall be no structural damage or deflection exceeding $L/250$, where L is the length of the side wall.

Integral supports provided with a tank must be constructed in accordance with the standard or subjected to the tank support load test. During this test, the tank and supports shall withstand a load of two times the weight of the full tank without evidence of permanent deformation or damage.

During production, each tank is subjected to the leakage test. Also, each dike wall is inspected for welding defects with a dye penetrant or other nondestructive testing method [2].

III. UL 80 AND UL 433

UL 80 was first published in 1927 (actually as Subject 142A) and has undergone minor changes over more than seven decades of use [3]. Because of the recent inception of higher-pressure filling practices, UL is considering increasing the strength requirements for UL 80 tanks. UL 433 has undergone minor changes since the first edition, which was published in May 1957 [4].

IV. UL 2085 AND SUBJECT 2080 OUTLINE OF INVESTIGATION

These documents cover aboveground tanks that include an insulation system intended to reduce the heat transfer to the primary tank should the construction be exposed to a hydrocarbon pool fire. The primary tank and the secondary containment are typically constructed to the requirements in UL 142. Glass fiber–reinforced primary and/or secondary containment tanks are covered when they comply with the same performance criteria.

The most significant performance difference between UL 2085 and Subject 2080 is the maximum allowable temperature on the primary tank during the full-scale fire test (Table 1).

Protected tanks, unlike fire-resistant tanks, must be subjected to a hose stream test immediately following the full-scale fire test. Both types of tanks have

Table 1 Full Scale Fire Test, Temperature Limits

	Subject 2080	UL 2085
Maximum average temp.	800°F average rise	260°F average rise
Maximum single-point temp.	1000°F	400°F

Source: UL 2085 and Subject 2080.

the option of being subjected to the vehicle impact test and ballistics (projectile) test. UL's listing mark identifies tanks that comply with these requirements.

UL 2085 was first published as an Outline of Investigation for Insulated Aboveground Tanks for Flammable and Combustible Liquids in November 1992. The test criteria were developed with the input of regulatory authorities and manufacturers. Insulated tanks, also referred to as fire-resistant tanks, were intended for installation in accordance with NFPA 30 (Flammable and Combustible Liquids Code) and NFPA 30A (Automotive and Marine Service Station Code).

In July 1993, the Uniform Fire Code (UFC) adopted new requirements for protected tanks. Following adoption, manufacturers of fire-resistant tanks requested that UL evaluate products based upon the UFC requirements. In March 1994, UL proposed that the new requirements for protected tanks be added to the Outline of Investigation for Insulated Tanks and further proposed that the outline be published as standard UL 2085.

In December 1994, UL published the first edition of UL 2085 (Standard for Insulated Aboveground Tanks for Flammable and Combustible Liquids). UL 2085 covers both types of tanks—fire-resistant tanks meeting NFPA 30 and NFPA 30A mandates, and protected tanks for UFC's requirements. Subsequently, UL transferred the requirements for protected tanks and fire-resistant tanks into the following documents: Subject 2080, which was published on December 29, 1997, and UL 2085, under the revised name of Standard for Protected Aboveground Tanks for Flammable and Combustible Liquids, published December 30, 1997 [5].

V. UL 2244

This document covers factory-fabricated, preengineered aboveground tank systems. These systems include a primary tank, integral secondary containment, breathing vents, emergency vents, overfill prevention systems, liquid level gauges, piping, antisiphon valves, access ladders, and other components required by installation codes. Listed tank systems are provided with a code compliance verification list (CCVL) which documents adherence to AST installation codes, including NFPA 30, NFPA 30A and the Uniform Fire Code. The CCVL addresses supports, venting, piping and fittings, tank construction, electrical installations, spill control features, dispensers, and environmental regulations. The regulatory authority determines whether the compliance assumptions documented on the CCVL are appropriate for specific tank-system installations.

The UL listing mark on the tank system identifies the construction standard of the primary tank. For example, UL 142/UL 2244 indicates a steel tank system, while UL 2085/UL 2244 indicates a protected tank system [6]. The marking also identifies the intended use of the tank system. The following types of systems are covered under UL 2244.

A. Part I—Motor Vehicle Fuel Dispensing Tank Systems

These systems are intended for storing Class I, II, or III-A liquids for refueling motor vehicles. Dispensers and fill connections are located on top of the tank, remote or on the side. Remote field-installed sumps, piping, wiring, islands, etc., are not evaluated as part of the system.

B. Part II—Generator Base Tank Systems

These systems are intended for storing Class II and III liquids used as a source of fuel for standby and emergency power generators that are mounted on top of the storage tank.

C. Part III—Aviation Fuel Storage Tank Systems

These systems are intended for storing Class I, II, and III-A aviation fuels, which are dispensed into aircraft-refueling vehicles and/or directly into aircraft.

D. Part IV—Motor Oil Storage Tank Systems

These systems are intended for storing Class III-B liquids such as motor oil or used motor oil (crankcase drainings).

VI. CONCLUSION

During the last decade UL has responded to significant changes in the storage tank industry, particularly in the development of new AST designs. As technology continues to evolve—in conjunction with public policy—UL standards will focus on ensuring that the latest innovations provide adequate safety for future generations.

REFERENCES

1. Underwriters Laboratories. Protected Aboveground Tanks for Flammable and Combustible Liquids, UL 2085. 1997.
2. Underwriters Laboratories. Steel Aboveground Tanks for Flammable and Combustible Liquids, UL 142, 1993.
3. Underwriters Laboratories. Steel Inside Tanks for Oil Burner Fuel, UL 80. 1996.
4. Underwriters Laboratories. Steel Auxiliary Tanks for Oil Burner Fuel, UL 433, 1995.
5. Underwriters Laboratories. Outline of Investigation for Fire Resistant Aboveground Tanks for Flammable and Combustible Liquids, Subject 2080. 1997.
6. Underwriters Laboratories. Aboveground Tank Systems for Flammable Liquids, UL 2244. 1999.

19

Development of ULC Standards for Aboveground Storage and Handling of Flammable and Combustible Liquids

Gordana Nikolic
Underwriters' Laboratories of Canada, Scarborough, Ontario, Canada

I. INTRODUCTION

As noted in Chapter 10, Underwriters' Laboratories of Canada (ULC) is a not-for-profit organization with headquarters in Toronto, (Scarborough), Ontario, Canada. It is a Canadian safety, certification, testing, quality registration, and standards development organization. ULC has been certifying and developing standards for flammable and combustible liquid storage products for more than 50 years.

Increased environmental concern produced the growth in regulation of underground storage tanks. The result has led to a marketplace trend to replace the traditional underground storage with aboveground systems, where the tank is readily available for inspection and any leakage can be detected visually before significant quantities of product could be released to the environment.

Environmental requirements for tanks often parallel fire protection concerns, and inevitably become interwoven with them. However, on occasion such requirements may be in conflict. A typical example is one of the newer requirements that has emerged—secondary containment for aboveground storage tanks. Secondary containment is designed to prevent leakage to the environment, but the retention of spilled or leaked flammable liquids in an open container presents a fire risk.

II. STANDARDS AND OTHER RECOGNIZED DOCUMENTS

With the increase in the popularity of aboveground storage tanks, regulators demanded protection from the risks inherent with the use of aboveground storage tanks. In response, the Steel Tank Committee has produced a number of standards for aboveground tanks and related products. The following standards have been published.

- ULC-S601 (Shop Fabricated Steel Aboveground Horizontal Tanks for Flammable and Combustible Liquids)
- CAN/ULC-S602 (Aboveground Steel Tanks for Fuel Oil and Lubricating Oil)
- ULC-S630 (Shop Fabricated Steel Aboveground Vertical Tanks for Flammable and Combustible Liquids)
- CAN/ULC-S643 (Shop Fabricated Steel Aboveground Utility Tanks for Flammable and Combustible Liquids)
- ULC-S652 (Tank Assemblies for Collection of Used Oil)
- ULC-S653 (Aboveground Steel Contained Tank Assemblies for Flammable and Combustible Liquids)
- ULC-S655 (Aboveground Protected Tank Assemblies for Flammable and Combustible Liquids)

As it has done with underground tanks, ULC, under its accreditation by the Standards Council of Canada (SCC), has developed and published Other Recognized Documents (ORDs) to establish certification criteria. These ORDs are circulated to relevant regulatory authorities for approval and adoption. As represented on this list, ULC has prepared a number of ORDs for aboveground tanks and is developing new ORDs for specialized tanks and accessories:

- ULC/ORD-C142.5 (Concrete Encased Steel Aboveground Tank Assemblies for Flammable and Combustible Liquids)
- ULC/ORD-C142.6 (Storage Vaults)
- ULC/ORD-C142.13 (Mobile Refueling Tanks)
- ULC/ORD-C142.15 (Concrete Tanks)
- ULC/ORD-C142.17 (Special Purpose Relocatable Aboveground Vertical Tanks)
- ULC/ORD-C142.18 (Rectangular Steel Aboveground Tanks for Flammable and Combustible Liquids)
- ULC/ORD-C142.20 (Secondary Containment for Aboveground Flammable and Combustible Liquids Storage Tanks)
- ULC/ORD-C142.21 (Aboveground Used-Oil Systems)
- ULC/ORD-C142.22 (Contained Vertical Steel Aboveground Tank Assemblies for Flammable and Combustible Liquids)
- ULC/ORD-C142.23 (Aboveground Waste Oil Tanks)

In the fields of leak prevention and detection, ULC was requested by the National Task Force of the Canadian Council of Ministers of the Environment (CCME) to develop a series of requirements reflecting environmental safety. Several of these can apply to aboveground tanks and their associated piping systems:

ULC/ORD-C58.9 (Secondary Containment Liners for Underground and Aboveground Flammable and Combustible Liquid Tanks)
ULC/ORD-C58.12 (Leak Detection Devices [Volumetric Type] for Underground Flammable Liquid Storage Tanks)
ULC/ORD-C58.14 (Non-Volumetric Leak Detection Devices for Underground Flammable Liquid Storage Tanks)
ULC/ORD-C58.15 (Overfill Protection Devices for Flammable Liquid Storage Tanks)
ULC/ORD-C58.19 (Spill Containment Devices for Underground Flammable and Combustible Liquid Storage Tanks)
ULC/ORD-C107.4 (Ducted Flexible Underground Piping Systems for Flammable and Combustible Liquids)
ULC/ORD-C107.7 (Glass-Fibre Reinforced Plastic Pipe and Fittings for Flammable and Combustible Liquids)
ULC/ORD-C107.12 (Line Leak Detection Devices for Flammable Liquid Piping)
ULC/ORD-C107.19 (Secondary Containment of Underground Piping for Flammable and Combustible Liquids)
ULC/ORD-C107.21 (Under Dispenser Sumps)
ULC/ORD-C107.14 (Nonmetallic Pipe and Fittings for Flammable Liquids)
ULC/ORD-C142.19 (Spill Containment Devices for Aboveground Flammable and Combustible Liquid Storage Tanks)
ULC/ORD-C180 (Liquid Level Gauges and Indicators for Fuel Oil and Lubricating Oil Tanks)
ULC/ORD-C586 (Flexible Metallic Hose)

Because ORDs are approved and adopted by Canadian authorities, they are accepted and enforced throughout the country. Many of the ORDs and standards previously listed have also been incorporated into national and provincial codes and other regulations.

III. ABOVEGROUND STORAGE TANKS

A. General

Aboveground storage tanks are either stationary or relocatable nonpressure tanks, and are intended for the storage and handling of flammable or combustible liquids, such as gasoline, fuel oil, or similar products with a relative density not greater than 1.0. These products are listed under ULC's label service program. After the certification investigation is successfully completed, and a listing promulgated, periodic examinations and tests are conducted on samples selected at random from current production and stock.

Currently, ULC has nine listing categories for aboveground tanks (Table 1). The remainder of this section addresses specific aboveground tank types and their associated standards and ORDs.

B. Installation and Use of Aboveground Tanks for Flammable and Combustible Liquids

These products are intended for use aboveground and are installed in accordance with requirements of the:

National Fire Code of Canada
Environmental Code of Practice for Aboveground Storage Tank Systems Containing Petroleum Products
Environmental Code of Practice for Underground Storage Tank Systems Containing Petroleum Products and Allied Petroleum Products
CSA B139 (Installation Code for Oil Burning Equipment)
National Fire Protection Association code, NFPA 30 (Flammable and Combustible Liquids Code)
Requirements identified by authorities having jurisdiction (AHJ)

C. Aboveground Tank Standards

Published in August 1955, ULC's first set of requirements addressed both aboveground horizontal and vertical tanks. In 1984, two standards were published—ULC-S601, standard for horizontal tanks, and ULC-S630, standard for vertical tanks. The following subsections address the most common designs of the aboveground tanks.

1. ULC-S601 (Standard for Shop-Fabricated Steel Aboveground Horizontal Tanks for Flammable and Combustible Liquids)

This standard covers horizontally oriented storage tanks of the non-pressure type that are intended for aboveground storage of flammable and combustible liquids such as gasoline, fuel oil, and similar products. The requirements provide construction-based information, such as minimum steel thickness, maximum diameter, diameter-to-length ratio, weld joints, and minimum vent openings.

These tanks may be of single-wall, double-wall, or compartmentalized construction. Double-wall tanks include a secondary containment that covers a minimum of 300° of the circumferential surface area of the primary tank and 100 percent of the primary tank heads. Construction of the secondary containment is separate from, but includes attachment to, the primary tank by stitch welding along the circumferential edges of each shell plate. This provides an interstice between the primary tank and secondary containment. The remaining area at the top of the tank is single wall where fittings are normally located.

All primary tanks are equipped with normal emergency venting and may also include a manhole access.

All single- and double-wall tanks must be leakage tested by applying pressure to the primary tanks as specified in the ULC-S601 standard.

Double-wall tanks must have emergency pressure relief for the interstitial space. They are leak tested by drawing required vacuum on the interstice and are factory equipped with permanent vacuum-monitoring devices. These tanks are intended for stationary aboveground installation only.

2. ULC-S630 (Standard for Shop Fabricated Steel Aboveground Vertical Tanks for Flammable and Combustible Liquids)

This standard addresses vertically oriented storage tanks of the nonpressure type that are intended for the aboveground storage of flammable and combustible liquids such as gasoline, oil, and similar products. The requirements provide construction-based design information, such as minimum steel thickness, maximum diameter, diameter-to-length ratio, weld joints, and minimum vent openings.

These tanks may be of single- or double-wall construction. The secondary containment shall envelope the bottom and the side of the primary tank—up to a height 50 mm (2 in.) below the top of the tank. The bottom of the secondary containment shall be constructed with material of equivalent or greater thickness than the primary tank.

Construction of the secondary containment is separate from, but includes attachment to, the primary tank by stitch welding along the circumferential edges of each shell plate. This, as described previously, provides an interstice.

All primary tanks must have provisions for normal and emergency venting and may include a manhole access.

All single- and double-wall tanks are required to be leakage tested by applying pressure to the primary tanks as specified in the ULC-S630 standard.

Double-wall tanks must have emergency pressure relief for the interstitial space. Factory equipped with permanent vacuum monitoring devices, all double-wall tanks are required to be leak tested by drawing the required vacuum on the interstice.

Tanks equipped with the fill pipe connection at the top of the tank are provided with a spill containment device of not less than 15 L capacity. This will meet

Table 1 Canadian Aboveground Tank Standards

Standard/document	Type	Description	Comment
	Aboveground tanks (ULC Guide No. 60 O5.0)		For Øammable and combustible liquids
ULC-S601	Horizontal aboveground steel tank	Single, double, compartment	
ULC-S630	Vertical aboveground steel tank	Single, double	
ULC/ORD-C142.18	Rectangular aboveground steel tank	Single, double	
	Fuel oil tanks (ULC Guide No. 60 O5.05)		For new and used fuel oil supply tank for oil-burning equipment
CAN/ULC-S602	Steel tank for fuel oil	Cylindrical, obround, rectangular, single, double	
	Utility tanks (ULC Guide No. 60 O5.01)		For Øammable and combustible liquids
CAL/ULC-S643	Utility stell horizontal tank	Single, double	
	Tanks for used oil (ULC Guide No. 60 O5.03)		For collection of used oil
ULC-S652	Tank for collection of used oil	Single, double, contained, encased,	ModiÆed aboveground and underground tanks
ULC/ORD-C142.23	Tank for manual deposition of used oil	vertical, contained, c/w drum, pump, monitor, tank workbench	ModiÆed S630 tanks
ULC/ORD-C142.21	Used oil system		Drum collection unit
ULC/ORD-C142.18	Rectangular aboveground steel tank		
	Contained tank assemblies (ULC Guide No. 60 O5.0.3)		For Øammable and combustible liquids
ULC-S653	Contained horizontal tank assembly	Single, double, containment	S601, S602, S643 primary tanks
ULC/ORD-C142.22	Contained vertical tank assembly	S630 primary tank	
ULC/ORD-C142.18	Contained rectangular tank assembly	Single, double, compartment	Rectangular primary tanks
ULC/ORD-C142.20	Secondary containment	RetroÆt/tanks up to 5000 L	

	Concrete encased tank assemblies (ULC Guide No. 60 O5.0.5)		For Øammable and combustible liquids
ULC/ORD-C142.5	Concrete-encased aboveground tank assembly	Single, double, compartment	
	Protected tank assemblies (ULC Guide No. 60 O5.0.16)		For Øammable and combustible liquids
ULC-S655	Protected aboveground tank assembly	Single, double, compartment	
	Mobile refueling tanks (ULC Guide No. 60 O5.0.20)		
ULC/ORD-C142.13	Mobile refueling tanks	Single, double, contained	
ULC/ORD-C142.17	Relocatable aboveground vertical tanks		
	Concrete tanks[a,b]		OK for the following applications:
ULC/ORD-C142.15	Concrete tanks	Not suitable for Øammable and combustible liquids	Oil-water separator Secondary containment Protected tanks Septic tanks and similar products
ULC/ORD-C142.6	Storage vaults[a]		
	Mainly for the storage and handling of Øammable and combustible liquids		
	Accessories, Øammable liquid tanks (ULC Guide No. 60 O5.20)		
ULC/ORD-C58.9	Secondary containment liners		For underground and aboveground Øammable liquid tanks

[a]Under development.
[b]Guide no. to be assigned.
Source: Underwriters' Laboratories of Canada.

the requirements of ULC/ORD-C142.19 (Spill Containment Devices for Aboveground Flammable and Combustible Liquid Storage Tanks). Such tanks also must have an overfill protection device that will meet the requirements of ULC/ORD-C58.15 (Overfill Protection Devices for Flammable Liquid Storage Tanks). These tanks are intended for stationary aboveground installation only.

3. CAN/ULC-S643 (Standard for Shop Fabricated Steel Aboveground Utility Tanks for Flammable and Combustible Liquids)

This standard addresses the construction and performance of steel aboveground tanks of the nonpressure type that are limited to a maximum of 5000 L capacity. They are of cylindrical horizontal design and equipped with integral support to allow for relocation. They are commonly used for:

Temporary fuel supply at construction sites
Permanent field installations supplying farm equipment with diesel
Private outlets with limited fuel requirements

Traditionally, the majority of these tanks have been single wall, but environmental concerns have led to the development of a double-wall version of the utility tank. Consequently, the standard's latest edition covers requirements for both single- and double-wall constructions.

This standard contains construction and performance criteria. It specifies criteria such as minimum steel thickness and weld joints in addition to performance tests that demonstrate the strength of the tanks. The tests include a hydrostatic pressure test, tank drop test and lifting lug strength test. After being subjected to these tests, the tanks shall be leak tight without permanent damage or distortion.

4. ULC/ORD-C142.18 (Rectangular Steel Aboveground Tanks for Flammable and Combustible Liquids)

The requirements for rectangular tanks address both construction and performance criteria and cover single-wall steel tanks and tanks with secondary containment (either double walled or contained). This standard addresses:

Workbench tanks, for combination use as a working surface and reservoir of lubricating oil
Generator base tanks that are integrated into the support structure of a diesel engine generator and used for storage of the generator's fuel supply

A hydrostatic pressure test, tank drop test, and lifting lug strength test are part of the performance criteria. Additionally, tanks are subjected to the static-load test during which a 1000-ob load is placed on the weakest area of the tank's top surface. After these tests, tanks are examined for leakage, damage, or distortion. A vacuum test is also required for double- wall tanks.

All rectangular tanks are subjected to the support load test. During this test, the tank and support must withstand a load of two times the weight of the full tank without evidence of permanent deformation or damage. Generator base tanks are subjected to twice the generator load or twice the load specified by the manufacturer.

It is important to note that bench tanks are required to have several warnings regarding their use. For example, welding is not permitted on top of a bench tank.

5. Aboveground Steel Contained Tank Assemblies for Flammable and Combustible Liquids

The initial contained horizontal aboveground tank ORD, published in 1992, addressed construction of the tank assembly. This consisted of a ULC-S601 horizontal primary tank contained in a rectangular closed box with a capacity equal to or greater than 110 percent of the primary tank. The 110 percent capacity was drawn directly from the National Fire Code. It represents a reasonable safety margin to address the accumulation of precipitation that could occur in an open dike. This was the first set of requirements developed to deal with minimum specifications for the complete tank assembly:

The tank
Spill containment
Overfill protection devices
Antisiphon valves
Stairs for manual filling where the distance between the receptacle and grade exceeds 1 m (3.25 ft) in height

In 1995, this ORD was upgraded to a standard, ULC-S653 (Aboveground Steel Contained Tank Assemblies for Flammable and Combustible Liquids).

In the published standard, the committee changed the requirements to allow acceptance of primary tanks that meet established ULC tank requirements in addition to the ULC-S601 Standard. (Previously only tanks meeting the ULC-S601 Standard were accepted.) The only restriction on this is that the primary tank cannot be vertical. This allowed various designs—some watertight and some airtight—to be used. Consequently, ULC/ORD-C142.22 (Contained Vertical Steel Aboveground Tank Assemblies for Flammable and Combustible Liquids) was published in 1995 to cover requirements for vertical tanks.

Both ULC-S653 and ULC/ORD-C142.22 are performance-based standards, requiring a product to pass a leakage test, a hydrostatic load test and a buoyancy test. For the buoyancy and hydrostatic load test, the containment area surrounding an empty primary tank is filled with water to its maximum capacity. The tank is not permitted to float within the containment, which is examined for leakage, damage, or permanent distortion. Other required tests for the assemblies are contained in the standards for the individual components.

6. Protected Tank Assemblies

ULC/ORD-C142.5 (Concrete Encased Steel Aboveground Tank Assemblies) is the pioneer document for tanks having thermal insulation protection. A two-hour pool fire test is performed to determine whether the product within a primary tank is protected from heat caused by an external fire condition. The design provides a containment that is capable of accommodating most forms of leakage from primary tanks, and affords a degree of protection against external fire. The primary tank must meet the requirements of either ULC-S601 or ULC-S643.

The next generation of requirements resulted from a need to evaluate tanks intended to be protected from external fire, vehicle impact, and small arms attack. The Canadian (ORD-C142.16) and American (UL 2085) protected tanks standards were developed simultaneously. Both ULC and UL adopted already established furnace test requirements for construction materials. It was also an easier test to control, compared to the pool fire test. The simulated fire test is conducted in a furnace where the heat is controlled on a rapid-rise temperature curve. This test requires a rapid temperature rise from ambient to 1093°C in five minutes to simulate a hydrocarbon pool fire; thereafter, the achieved temperature is maintained for two hours. To meet ULC requirements, the assemblies must also undergo a simulated vehicular impact test and bullet resistance test. Protected tank assemblies provide secondary containment that is capable of containing most forms of leakage from the tanks and, in addition, an encasement that affords a degree of protection against vehicle impact, small arms attack, and external fire.

ULC's standard committee upgraded the ORD into a national standard. The first edition of standard ULC-S655 (Aboveground Protected Tank Assemblies for Flammable and Combustible Liquids) was published in December 1998.

IV. CONCLUSION

ULC's challenge as a standards-writing organization is to find environmentally safe and cost-effective approaches in developing product requirements. One of our core values is a concern about safety: both fire and environmental. A significant portion of today's development of our standards in the area of flammable and combustible liquids is tied to environmental protection performance. The ULC and cUL mark on a product confirms compliance with established Canadian standards or ORDs.

20
Secondary Containment for Noninsulated Steel Aboveground Storage Tank Systems

Jerry L. Waller
Modern Welding Company, Inc., Owensboro, Kentucky

I. ORIGINS OF THE SECONDARY CONTAINMENT AST

During the early 1980s, government regulators had become increasingly concerned with protecting the environment. Contaminants from underground petroleum tank systems were the regulators' main focus. Rumors surfaced that new underground storage tank (UST) regulations were in development, which might include financial responsibility requirements for tank owners. The tighter regulation of USTs opened a new era in the storage tank industry. It was the dawn of a movement toward the use of aboveground storage and new nationally available, environmentally sound tank technologies.

II. DIKED TANK DESIGN

A few steel tank manufacturers saw the opportunities presented by the regulations and started to fabricate a small-capacity (2000 gal and under) aboveground storage tank (AST) system with secondary containment—that is, the diked tank, or a tank in a box. The diked tank became the first shop-fabricated aboveground tank manufactured with secondary containment to catch spills and overfills of the tank. The diked tank, featuring a unitized, lighter-weight construction, used a UL 142 cylindrical steel tank set inside an open-top steel box (Fig. 1). This made it easy

Figure 1 Open top diked tank meeting STI F911 specification.

to relocate, unlike conventional dike systems. This standard has been used primarily for stationary tanks under a 2000-gal capacity, though some have been fabricated with capacities as large as 50,000 gal.

In 1991, this type of secondary containment system became a Steel Tank Institute standard (F911). Around the same time, it received an Underwriters Laboratories (UL) listing.

This technology provided significant environmental security because of the impervious quality of the steel secondary containment and the ease of visually inspecting the tank and the inside and outside of the containment area. The technology also had some drawbacks. The secondary containment was not sealed from the weather, which enabled precipitation to enter the diked area. This had three significant impacts:

The secondary containment area was required to hold 110 percent capacity of the primary single-wall tank because of the potential for capturing precipitation in the diked area. This meant a larger area for installation was required, increased shipping costs.

Liquids had to be removed to maintain the secondary containment capacity.

Rainwater or snow could become mixed with small amounts of petroleum, transforming the liquid into a hazardous waste the disposal of which was costly.

The combination of an open-top design and precipitation created another significant issue. The F911 standard allowed drain openings in the dike, which also led to environmental concerns. When water had to be removed from the dike, one option was to use the sidewall or bottom drain. If not properly safeguarded, this could have resulted in a release into the environment. The most environmentally protective method of liquid removal was to specify a dike without drain holes, which required that the tank manager make provisions for pumping excess water out of the secondary containment.

The F911 has been primarily used for small storage applications that aren't moved often. Improvements such as overflow piping and rain shields have been added. Rain shields, or closed-top designs, prevent precipitation from entering the secondary containment area (Fig. 2).

III. DOUBLE-WALL DESIGN

The next step in secondary containment for shop-fabricated ASTs was a design that, in essence, created two steel tanks in one. During the mid-1980s, a large machine shop wanted an aboveground double-walled tank built. The tank was to store used cutting fluids from the shop's machining operations. The machine shop had been using 55-gal drums for storing the fluids; spills from the drums would create

Figure 2 Rain shield design. Photo supplied by Modern Welding.

cleanup problems because the used cutting fluids were considered hazardous waste.

The machine shop owner stipulated a few requirements for the aboveground double-walled tank:

- Weather-tight secondary containment
- A design that would hold at least 100 percent of the liquid stored in the primary tank
- Capability to monitor the space between the inner and outer tank
- A vertical design because of space limitations
- Emergency venting in the primary and secondary tanks

The first tank was built and installed with a continuous monitoring device for leak detection between the inner and outer tank. This tank, which was unprecedented in design and function, is now known as the vertical F921 (Fig. 3).

Some thought that this type of tank might be beneficial for environmentally safe aboveground fuel storage. Market surveys found a need for water-tight aboveground tanks that provided secondary containment, could be monitored and required a smaller installation area.

By 1988, a complete line of horizontal and vertical double-walled aboveground tanks had been developed. The designs received a UL listing to store flammable and combustible liquids. Sales took off as customers recognized design characteristics that addressed marketplace concerns.

Water-tight construction eliminated the pumping of fluids from the secondary containment area. This allowed the secondary containment capacity to be decreased to 100 percent of the product contained in the primary tank, which helped to reduce shipping costs.

Intimate-wrap construction was lighter weight, allowed vacuum testing for leaks, and reduced the installation area. It also decreased long-term maintenance because the total painted area exposed to the elements was 50 percent less than the F911 technology. The F921 standard allowed easier relocation to future sites than any other AST secondary containment system.

As the popularity of double-walled aboveground technology grew, the Steel Tank Institute began work to create a new national standard. In 1992 the Steel Tank Institute published the new F921 national double-wall vertical and horizontal aboveground storage tank standard (Figs. 3, 4). This new technology's primary use has been for the storage of fuel oil for industrial use and fleet fueling applications.

IV. OPERATIONAL ISSUES

The most common sizes for F921 tanks have been 8,000, 10,000 and 12,000 gal. Tank owners typically ordered within this range of capacities because they can ac-

Figure 3 Double wall vertical aboveground storage tank with interstitial monitoring pipe.

Figure 4 Horizontal F921 design with interstitial monitoring pipe on tank head.

commodate a full transport drop of fuel. Tanks of 12,000 gal and larger have often been compartmentalized to allow for multiproduct storage in one tank. The F921 tanks have been built from 50 to 50,000 gal capacities.

Fire codes require dikes for ASTs with greater than 12,000 gal capacity. However, local authorities have final jurisdiction on all installations. They may restrict location, use, and sizes allowed.

In April of 1993 Underwriters Laboratories published the seventh edition of the UL 142 standard, which incorporated the concept of the Steel Tank Institute's F911 and F921 standards.

During 1998, the Uniform Fire Code adopted language that allows dispensing of motor fuel into vehicles at locations not open to the public from F921 tanks. Prior to the change, the UFC would only permit such dispensing from fire-rated tanks. Nevertheless, in some areas of the country covered by the Uniform Fire Code, local fire marshals prior to 1998 would allow the use of non fire-rated tanks on an individual basis. The areas covered by National Fire Protection Association

(NFPA) standards for several years have allowed the dispensing of motor fuels from non fire-rated tanks.

A decade after the introduction of the secondary containment AST designs, sales and popularity have continued to grow in the United States and around the world.

21
AST Environmental Regulations

Wayne B. Geyer
Steel Tank Institute, Lake Zurich, Illinois

I. BACKGROUND

To operate an aboveground storage tank is to invite regulatory scrutiny. Fire and building codes have changed dramatically during recent years to address rapidly shifting marketplace demand. But significant environmental protection rules also must be considered.

The most prominent federal environmental program to regulate aboveground storage tanks (ASTs) is the Spill Prevention Control and Countermeasures (SPCC) effort. However, on several occasions during the last decade, various members of Congress have introduced legislation that would intensify AST requirements. For a number of reasons, none of the legislative initiatives have reached critical mass.

However, many state, county, and local environmental authorities have also enacted policies that govern the usage of ASTs. A quick survey of environmental regulators prior to the start of an aboveground storage project will save time later trying to fix issues that could have been easily anticipated.

II. FEDERAL REGULATIONS

The U.S. Environmental Protection Agency (EPA) regulates aboveground storage tanks through two pieces of legislation—the Clean Water Act (33 USC 1251 et seq.), and the Oil Pollution Act of 1990 (33 USC 2701 et seq.) EPA is responsible for, among other things, protecting the nation's waters from the adverse effects of oil spills. The SPCC regulation, which enacts section 311(j) of the Clean Water Act, is designed to prevent discharges of oil from storage facilities and to contain such releases when they occur.

The regulation applies to "onshore, non-transportation-related facilities" that could reasonably be expected to discharge oil into navigable waters, when such facilities have:

An aboveground oil storage capacity of more than 660 gal in a single container
A total aboveground oil storage capacity of more than 1320 gal in multiple containers
A total underground oil storage capacity of more than 42,000 gal [1].

According to EPA, non-transportation-related facilities are all fixed facilities, including support equipment, but excluding certain pipelines, railroad tank cars en route, transport trucks en route, and equipment associated with the transfer of bulk oil to or from water transportation vessels. The term also includes mobile or portable facilities, such as drilling or work-over rigs, production facilities, and portable fueling facilities while in a fixed, operating mode [2].

There is no generally accepted statistical measure of how many tanks can be affected by SPCC requirements. For instance:

The Petroleum Marketers Association of America (PMAA) reported in 1997 that its members operated 10,000 bulk storage facilities, which, of course, could include multiple tanks. PMAA also reported that a typical association member had five aboveground storage tanks at a bulk facility with 125,000 gal aggregate storage capacity [3].
The American Petroleum Institute (API) in 1989 published a survey that estimated about 700,000 ASTs were used nationally in marketing, refining, transportation, and production segments of the petroleum industry [4].
The Environmental Defense Fund expanded API's projection to include between 100,000 and 200,000 smaller distribution facilities that may have been overlooked by the 1989 estimate [5].
A 1996 EPA survey said that about 438,000 storage facilities are potentially covered under the SPCC regulation [6].

EPA also estimates that between 70 and 80 percent of SPCC-regulated tanks are used in two industries (farms and oil production). Though farms are one of the largest categories of storage facilities, only about 8 percent of farms utilize tanks that would be regulated by SPCC rules. The remaining SPCC tanks are employed in manufacturing, transportation, gasoline stations, vehicle fueling facilities, or other industries.

EPA's data indicate that facilities in different industry sectors vary dramatically in total storage capacity, number of tanks, and annual throughput volume. Farms in general have smaller storage capacity, fewer tanks, and lower throughput levels than other types of facilities.

The 1996 survey investigated whether statistically significant relationships exist between the characteristics of oil storage facilities and their propensity to spill oil. The results showed that facilities with greater oil storage capacity are likely to have a greater number of oil spills, larger volumes of oil spilled, and greater cleanup costs than facilities with smaller capacity. EPA concluded that as total storage capacity, number of tanks, and annual throughput increase, so do the propensity to spill, the severity of spills, and the attendant costs of cleanup.

EPA's analysis revealed that SPCC compliance reduces the number of spills, spill volume and the amount of oil that migrates outside of a facility's boundaries. The survey also revealed that a large proportion of facilities that meet the SPCC regulatory requirements might not be in full compliance [1].

EPA also faces some pressure to lessen the risk of oil spills by promoting the use of standards for tank construction and testing, and strengthening AST inspection procedures. A July 1995 report from the General Accounting Office said EPA should require "that ASTs be built and tested in accordance with the industry's or other specified standards." The GAO was echoing its recommendation from 1989, which followed some major oil spills from field-erected tanks. EPA officials told GAO in 1995 that the agency wanted to propose simultaneously rules on tank construction, contingency plans and inspections.

EPA had proposed rules in 1991 that would require integrity testing of the tank every five years unless the AST facility featured secondary containment. An AST system with secondary containment would require integrity testing every 10 years and when major repairs are made. The proposed rules also would mandate AST facilities without secondary containment to conduct integrity and leak testing of valves and piping once a year at minimum [7]. The proposal also suggests that all secondary containment be impermeable for at least 72 hours.

A. What are the Problems with Regulating ASTs?

EPA in September 1993 conducted four forums to learn more about aboveground storage tank systems for petroleum. Stakeholders from all sectors of the storage tank industry participated in the meetings that were conducted in Philadelphia, Washington, DC, Austin, and San Francisco. Six themes emerged from the discussions:

The problem of soil and groundwater contamination through discharges and spills from AST facilities is not well documented, although evidence of a problem does exist.

AST facilities are diverse.

The AST population is growing.

Most contamination is from past practices and older tanks and facilities.

Causes and sources of leaks from tanks and piping vary.
Small facilities are responsible for much of the contamination from AST systems.

Furthermore, when the forums examined regulations and industry standards, the experts found:

The current system of regulations and industry standards for AST facilities is both confusing and inefficient.
The current system of legislation and regulations for AST facilities has gaps.
Outreach to support implementation of current regulations and enforcement of those regulations is weak.
The current system of regulations and industry codes strongly encourages environmental protection.
Costs of complying with current regulations are unequally distributed.
Certain regulatory criteria and/or requirements for AST facilities are technically difficult or impractical to implement [8].

One of the more challenging aspects of regulating ASTs during the 1990s was keeping up with industry innovation. Although the 1990s witnessed numerous unsuccessful attempts to pass new legislation that would further regulate ASTs, industry responded with a number of new standards to address environmental concerns. These include:

API 653 (Tank Inspection, Repair, Alteration, and Reconstruction) from American Petroleum Institute
API 2610 (Design, Construction, Operation, Maintenance, and Inspection of Terminal & Tank Facilities) from American Petroleum Institute; API 2610 compiled where and how all of API's standards for ASTs can be properly used to safely design, store, and operate AST systems
PEI/RP 200-96 (Recommended Practices for Installation of Aboveground Storage Systems for Motor Vehicle Fueling) from Petroleum Equipment Institute
F921 (Standard for Aboveground Tanks with Integral Secondary Containment) from Steel Tank Institute
F911 (Standard for Diked Aboveground Storage Tanks) from Steel Tank Institute

B. An Emerging Trend: USTs to ASTs

EPA also examined SPCC facilities during 1995 to determine whether ASTs were replacing underground storage tanks (USTs) at SPCC-regulated facilities. The EPA study, analyzing tank replacement activity between 1993 and 1995, showed that

5062 facilities had replaced a total of 27,462 USTs during the two-year period. The 27,462 USTs were replaced by 17,195 tanks, of which 56 percent (9,634 tanks) were ASTs and 44 percent (7561) were USTs [9].

C. Developing an SPCC Plan

The EPA's oil pollution prevention policies require that an SPCC plan be carefully thought out, prepared in accordance with good engineering practices, and approved by a person with the authority to carry out the plan. The SPCC plan must clearly address:

Operating procedures that prevent oil spills
Control measures installed to prevent a spill from reaching navigable waters
Countermeasures to contain, clean up, and mitigate the effects of an oil spill that reaches navigable waters

Each SPCC plan must be unique to the storage facility. Development of a unique SPCC plan requires detailed knowledge of the facility and the potential effects of any oil spill. Each SPCC plan, while unique to the facility it covers, must include certain standard elements to ensure regulatory compliance. These elements include:

Written descriptions of any spills occurring within the past year, corrective actions taken, and plans for preventing reoccurrence
A prediction of the direction, rate of flow, and total quantity of oil that could be discharged from any petroleum storage and handling equipment
A description of containment and/or diversionary structures or equipment to prevent discharged oil from reaching navigable waters
Where appropriate, a demonstration that containment and/or diversionary structures or equipment are not practical and a strong oil spill contingency plan and a written commitment of manpower, equipment, and materials to quickly control and remove spilled oil
A complete discussion of spill prevention and control applicable to the facility and/or its operations

The SPCC plan should include a demonstration of management's approval and should be certified by a registered professional engineer [10].

D. Amending an SPCC Plan

Owners or operators of SPCC facilities must amend the facility compliance plan whenever there is a change in design, construction, operation, or maintenance that materially affects the facility's potential to pollute navigable U.S. waters or adjoining shore lines. Amendments must be set in place as soon as possible but no later than six months after a change occurs.

Owners and operators also shall complete a review and evaluation of the SPCC plan at least once every three years from the date the facility qualified for SPCC regulation. As a result of this review and evaluation, the owner or operator shall amend the SPCC plan within six months of the review to include more effective prevention and control technology if the technology:

Will significantly reduce the likelihood of a spill event from the facility
Has been field-proven at the time of the review

A professional engineer must certify all SPCC plan amendments [11].

E. Facility Response Plans

Oil storage facilities that are governed by SPCC requirements—and could cause "substantial harm" to the environment—must prepare and submit a Facility Response Plan to EPA for review. A facility has the potential to cause substantial harm, according to EPA's definition, if:

1. It transfers oil over water to or from vessels and has a total oil storage capacity, including both aboveground storage tanks and underground storage tanks, of at least 42,000 gal.
2. Its total oil storage capacity, including both ASTs and USTs, is greater than or equal to 1 million gallons, and one of the following is true:
   a. The facility lacks secondary containment for each aboveground storage area sufficient to contain the capacity of the largest AST within each storage area plus freeboard to allow for precipitation.
   b. The facility is located near an environmentally sensitive area that would be affected by a discharge.
   c. The facility is located near a public drinking water intake that could be shut down by a discharge.
   d. The facility has had a reportable spill greater than or equal to 10,000 gal during the past five years.

Owners and operators of SPCC facilities that could create substantial harm to the environment are responsible for preparing and implementing plans, training, and drills for responding to:

A worst-case discharge of oil
A substantial threat of such a discharge
Discharges smaller than worst-case events [12]

III. STATE AND LOCAL REGULATIONS

Specifiers should check with environmental regulators representing state, county or municipal governments to see if additional requirements would affect an AST facility. For example, during 1998, Florida developed a new set of comprehensive rules for ASTs that placed significant emphasis upon the use of secondary containment. In addition, Florida added requirements for concrete diking to assure that cracks, or other signs of structural problems, would not undermine the containment.

A. California Resources Board

The California Air Resources Board (CARB) requires that aboveground storage systems in that state accommodate Stage I and Stage II vapor recovery to reduce emissions that can lead to air pollution.

Some states and metropolitan areas outside of California have adopted other CARB requirements to improve local air quality. If you are specifying an AST for a metro area that historically has been cited for high ozone or NOx levels, it's recommended that you check with any agencies that have authority to regulate potential sources of vapor emissions.

REFERENCES

1. U.S. Environmental Protection Agency [Background posted on the World Wide Web.], Washington, DC. Retrieved November 23, 1998, from the World Wide Web. http://www.epa.gov/oilspill/spcc/survq&d.pdf
2. U.S. Environmental Protection Agency [Background posted on the World Wide Web.], Washington, DC. Retrieved November 23, 1998, from the World Wide Web. http://www.epa.gov/oilspill/faqs/opprfaqs.htm
3. Journal of Petroleum Marketing. Industry Report 1997. April:31–32, 1997.
4. U.S. General Accounting Office. Report to Congress (B-261262). July:17, 1995.
5. U.S. General Accounting Office. Report to Congress (B-261262). July:18, 1995.
6. U.S. General Accounting Office. Report to Congress (B-261262). July:4–5, 1995.
7. U.S. Environmental Protection Agency. Analysis of the Number of Facilities Regulated by EPA's SPCC Program [Background posted on the World Wide Web.], Washington, DC. Retrieved November 23, 1998, from the World Wide Web. http://www.epa.gov/oilspill/spcc/pap_tpop.pdf
8. U.S. Environmental Protection Agency. Aboveground Oil Storage Facilities Workgroup. Highlights of Forums on Aboveground Oil Storage Facilities.
9. U.S. Environmental Protection Agency. Analysis of Trends in the Replacement of Underground Storage Tanks at SPCC-Regulated Facilities [Background posted on the

World Wide Web.], Washington, DC. Retrieved November 23, 1998, from the World Wide Web. http://www.epa.gov/oilspill/spcc/pap_u2a.pdfeb
10. U.S. Environmental Protection Agency. What SPCC Plans Must Include [Background posted on the World Wide Web.], Washington, DC. Retrieved November 23, 1998, from the World Wide Web. http://www.epa.gov/oilspill/spccmust.htm
11. U.S. Environmental Protection Agency. SPCC Plan Rule, Sec. 112.5 Amendments by Owners or Operators [Background posted on the World Wide Web.], Washington, DC. Retrieved November 23, 1998, from the World Wide Web. http://www.epa.gov/region09/waste/sfund/oilpp/spcc/1125.html
12. U.S. Environmental Protection Agency. Oil Spill Program Update, Vol. 1, No. 4. U.S. EPA's Oil Program Center Report. Washington, DC, 1998.

22

Plan Review and Inspection of Aboveground Storage Tanks

Scott A. Stookey
City of Austin Fire Department, Austin, Texas

I. INTRODUCTION

The approval of the authority having jurisdiction (AHJ) must be obtained before an aboveground storage tank is installed. Typically, this is the local fire marshal or fire inspector. Approvals from different agencies may be required. For example, some jurisdictions require that the fire, zoning, and environmental protection departments review the design for an aboveground storage tank (AST) before an approval is granted.

The design for an aboveground storage tank installation may need to be supervised by a registered professional engineer. The load-bearing capacity of the tank foundation, the integrity of the secondary containment, and the materials and methods used to construct the piping system should be reviewed by a professional engineer. Jurisdictions can require an engineer to seal the design drawings, specifications, and pipe test reports.

Consideration must be given to the location of the storage tank. Fire officials are granted wide latitude in deciding the location and the type of tank they may allow. Federal laws and local land use regulations can limit where a tank is located. The tank must be on a site that is accessible by the fire department and have a water supply for firefighting operations.

Fire codes and standards specify that aboveground storage tanks be provided with several safety controls and features. Design plans and specifications for a tank installation must address the provision of secondary containment, tank vents, overfill prevention—and protection of the tank against damage.

This chapter is divided into two parts: Plan Review and Inspection. The plan review section will assist designers and plans examiners by explaining some concerns to resolve before a tank can be installed. Some of these issues include land use considerations, tank vents, secondary containment and evaluating hazardous-location electrical equipment. The chapter's second section explains the important inspection issues for new aboveground storage tanks. Significant areas of concern include tank nameplates, emergency vents, functional tests of certain components, and integrity testing of the tank and piping. After an AST is installed, it should be inspected annually to determine its condition and operating status. A section dealing with AST maintenance inspections is included.

This chapter explains the essential steps for installing an aboveground storage tank based upon requirements of the model fire codes. Four model fire codes in the United States regulate the storage and use of flammable and combustible liquids. An effort is underway to consolidate these four codes into a single fire code—the International Fire Code—the first edition of which is scheduled for publication in 2000. The International Fire Code is a collaborative effort between the members of the Uniform Fire Code, Southern Fire Prevention Code, the Building Officials and Code Administrators (BOCA) Fire Prevention Code, regulated industries and concerned organizations. The International Fire Code will contain requirements derived from the language found in the three model fire codes, and reference most provisions found in National Fire Protection Association (NFPA) standards.

Aboveground storage tanks commonly store flammable and combustible liquids. Designating a liquid as "flammable" or "combustible" is based on its flash point and boiling-point temperatures. Flash point is only one of a number of properties that must be considered in assessing the overall flammability hazard of a material; regulatory authorities use the measurement to classify liquids as either flammable or combustible. Flammable liquids have a flash-point temperature lower than 100°F; combustible liquids have flash-point temperatures equal to or greater than 100°F. Testing of liquid flash points is apparatus dependent and may not represent the absolute minimum temperatures at which a material may emit flammable vapors. The classification system used in the U.S. model codes and standards is shown in Table 1.

II. PLAN REVIEW OF ABOVEGROUND STORAGE TANKS

A. Land Use Regulations

1. Model Fire Code Requirements

The draft International Fire Code, the 1997 Uniform Fire Code, and the 1997 Southern Fire Prevention Code authorize the fire official to limit the size, regulate the location, or prohibit the installation of aboveground storage tanks. One exam-

Table 1 Model Fire Codes ClassÆcation for Flammable and Combustible Liquids

ClassiÆcation	Flash point temperature	Boiling point temperature
Class I-A Øammable liquid	Less that 73°F	Less than 100°F
Class I-B Øammable liquid	Less than 73°F	Equal to or greater than 100°F
Class I-C Øammable liquid	Equal to or greater than 73°F and less than 100°F	Not applicable to classiÆcation
Class II combustible liquid	Equal to or greater than 100°F and less than 140°F	Not applicable to classiÆcation
Class III-A combustible liquid	Equal to or greater than 140°F and less than 200°F	Not applicable to classiÆcation
Class III-B combustible liquid	Equal to or greater than 200°F	Not applicable to classiÆcation

ple of such a land use provision is found in Uniform Fire Code Section 7902.2.2.1. This provision authorizes the fire chief to establish locations where Class I and II liquids can be stored. Class I and II liquids have a closed-cup flash point temperature lower than 140°F. Some jurisdictions may prohibit the installation of aboveground storage tanks. In other cases, the approval to install an aboveground storage tank may only be issued after sufficient information demonstrates to the fire chief that the AST safety risk is no greater than an underground storage tank. Jurisdictions may limit the size and location of tanks using the local zoning laws. For example, it may be acceptable to locate an aboveground storage tank in an industrial park, but siting it near a residential neighborhood or apartment community is probably unacceptable. Interpretations will vary between jurisdictions; therefore, the permit applicant should contact the fire official early in the tank-selection process. An initial discussion with the fire official should answer several key questions:

Does the jurisdiction allow aboveground storage tanks?
Is an aboveground storage tank allowed at the location in question?
Which aboveground storage tank types are acceptable?
What are the fire code requirements for aboveground storage tanks?
Are there other local or state laws or rules that regulate the installation and operation of an aboveground storage tank?

2. Zoning Laws

Local governments adopt zoning and land development laws. Zoning laws regulate the size, type, structure, nature, and use of land or buildings. These regulations will vary by community and may prohibit installing an AST or restrict its size and location based on the property's zoning. In certain cases, a public hearing before the local zoning or planning commission is required before a tank can be installed.

Figure 1 Siting an aboveground storage tank in a heavily populated, congested urban area may be unacceptable. This AST was temporarily installed. A code compliance inspection showed that the aboveground storage tank was not listed, was not equipped with an emergency vent, and lacked secondary containment.

The public hearing offers citizens a venue for explaining their concerns about the proposed installation. Design professionals should determine the land-use requirements for the community and how they will regulate the aboveground storage of hazardous materials.

3. Department of Housing and Urban Development Requirements

The U.S. Department of Housing and Urban Development (HUD) has regulations that restrict the location of aboveground storage tanks near areas where federal monies are used for housing and urban development. Siting an AST in an area where buildings were constructed, rehabilitated, modernized, or converted with HUD financial assistance requires evaluation to ensure that an adequate separation distance is provided between the tank and the area of concern. The regulation (Siting of HUD-Assisted Projects Near Hazardous Operations Handling Conventional Fuel or Chemicals of an Explosive or Flammable Nature) is found in Title 24, Part 51 of the Code of Federal Regulations. The regulation was established in 1984 under the Housing and Community Development Act of 1968, 1969, and 1974. The purpose of this particular regulation is to:

- Establish safety standards that can be used as a basis for calculating acceptable separation distances for HUD-assisted projects from specific, stationary, hazardous operations that store, handle, or process hazardous substances
- Alert those responsible for the siting of HUD-assisted projects to the inherent potential dangers when such projects are located in the vicinity of such hazardous operations
- Provide guidance for identifying those hazardous operations that are most prevalent
- Provide the technical guidance required to evaluate the degree of danger anticipated from explosion and thermal radiation (fire)
- Provide the technical guidance required to determine an acceptable separation distance from such hazards

The requirements do not consider the toxic effects of the hazardous material or its combustion byproducts.

The regulation defines "hazard" as "any stationary container which stores, handles, or processes hazardous substances of an explosive or fire-prone nature." The term "hazard" does not include pipelines installed underground or in accordance with the Federal Pipeline Safety Act. The regulation exempts tanks with a capacity of 100 gal or less when they contain common liquid fuels, such as gasoline and kerosene. Underground storage tanks (USTs) are excluded from the requirements. This exclusion also applies when tank trucks fill USTs at gasoline service stations. The regulation does not apply to underground or aboveground fuel-oil storage tanks. However, the regulation does apply to barges, ships, railroad tank cars, and tank trucks that are loaded or unloaded at a stationary AST. Table 2 lists

Table 2 SpeciÆc Hazardous Materials Regulated by 24 CFR 51.201

Acetic acid	Carbon disulÆde	Ethyl benzene	Methyl alcohol
Acetic anhydride	Cellosolve	Ethyl dichloride	Methyl amyl alcohol
Acetone	Cresols	Ethyl ether	Methyl cellosolve
Acrylonitrile	Crude oil (petroleum)	Gasoline	Methyl ethyl ketone
Amyl acetate		Heptane	Naphtha
Amyl alcohol	Cumene	Hexane	Pentane
Benzene	Cyclohexane	Isobutyl acetate	Propylene oxide
Butyl acetate	No. 2 diesel fuel	Isobutyl alcohol	Toluene
Butyl acrylate	Ethyl acetate	Isopropyl acetate	Vinyl acetate
Butyl alcohol	Ethyl acrylate	Isopropyl alcohol	Xylene
Carbon bisulÆde	Ethyl alcohol	Jet fuel and kerosene	

Source: Ref 1.

the specific petroleum products and chemicals defined as hazardous in 24 Code of Federal Regulations (CFR) 51.201.

The regulation requires an adequate separation distance (ASD) between the hazard and a HUD-assisted project. ASD is "the actual distance beyond which the explosion or combustion of a hazard is not likely to cause structures or individuals to be subjected to blast overpressure or thermal radiation flux levels in excess of the safety standards in 24 CFR 51.203." Aboveground storage tanks are designed to store materials at atmospheric pressure, so the regulation limits their evaluation to the fire threat. Thermal radiation levels (also referred to as "thermal-radiation fluxes") are measured and expressed in units of power per unit area of the item receiving the energy.

The thermal-radiation safety requirements in 24 CFR 51.203 are:

> The allowable thermal-radiation flux level at the building shall not exceed 10,000 BTU/ft^2/hr
>
> The allowable thermal-radiation flux level for outdoor, unprotected facilities or areas of congregation shall not exceed 450 BTU/ft^2/hr

Wooden structures, window drapes, and trees will generally ignite when exposed for a relatively long period of time to thermal-radiation flux levels of about 10,000 BTU/ft^2/hr. At that intensity, it takes about 15 to 20 min for a wooden structure to ignite. The 10,000 BTU/ft^2/hr value is based on the ignition of wooden structures on level terrain.

Human exposure to thermal-radiation levels of 1,500 BTU/ft^2/hr causes intolerable pain after 15 sec. Longer exposure results in blistering, permanent skin damage, and possible death. Considering this relatively short period, 15 sec, it is quite possible that persons with impaired mobility, children, or the elderly would not be able to take refuge in a timely manner. Consequently, the HUD requirements for unprotected outdoor areas where people congregate are limited to thermal-radiation flux of 450 BTU/ft^2/hr. Exposure to this level of thermal radiation for a prolonged period of time has a limited detrimental effect—the same as a bad sunburn. Unprotected outdoor areas such as parks, open space, or playgrounds must be located such a distance from the potential hazard so the radiation flux level will not exceed 450 BTU/ft^2/hr.

The ASD is based on the area of the dike around the tank. The dike area is the perimeter of an impounding space that forms a barrier to liquid flowing in an unintended direction. The dike area may serve one or many tanks. The regulation does not address the use of tanks with integral secondary containment. If a listed secondary containment tank is used, the dike area can be the tank outer shell. A more conservative approach for listed secondary containment tanks is to determine the ASD caused by a liquid pool fire due to offloading operations from a tank vehicle. Because offloading areas are normally not contained, the area of the spill will be greater. A larger spill causes a larger pool fire; therefore, the ASD will be greater. HUD recommends that a spill of the entire contents of a 10,000 gal. tank vehicle be considered when calculating an ASD for areas where vehicles are loaded or unloaded.

Calculation of the dike area is required to obtain the fire width, which for aboveground storage tanks is the diameter of the fireball or pool fire. The fire width used in this regulation is based on the square root of the dike area or:

$$\text{Fire width} = \sqrt{\text{Dike area}}$$

The fire width and dike areas are used to find the ASD from buildings and to protect persons from injuries. Table 3 specifies the acceptable separation distances for given dike areas [1]. For example, a manufacturing plant installs a 12,000-gal aboveground storage tank for isopropyl alcohol, which is a HUD-regulated hazardous material. The listed, single-wall tank is located inside a 400-ft^2 dike. A 400-ft^2 dike area has an estimated fire width of 20 ft. Using Table 3, the minimum required separation between the tank and any HUD-assisted building is 17 ft. If persons use the building, the required separation is 107 ft.

Acceptable separation distances are measured from the center of the tank [2]. If a facility has two or more dike areas, only the dike area closest to the HUD-assisted project requires evaluation. If a site has atmospheric storage tanks and pressure vessels containing regulated hazardous materials, HUD requires an evaluation of the storage tank closest to the HUD-assisted project and the pressure vessels, regardless of their location. In many instances, pressure vessels will have greater ASDs than atmospheric storage tanks.

Application of the standards for determining an acceptable separation distance between a HUD-assisted project and a potential hazard of an explosion or fire is predicated on level topography with no intervening object(s) between the hazard and the project. Application of the standards can be eliminated or modified if:

- The site topography shields the proposed project from the hazard
- An existing, permanent, fire-resistant structure of adequate size and strength will shield the proposed project from the hazard
- A barrier is constructed surrounding the hazard, at the site of the project or in between the potential hazard and the proposed project
- The structure and outdoor areas used by people are designed to withstand blast overpressure and thermal radiation anticipated from the potential hazard (e.g., the structure is of masonry and steel or reinforced concrete and steel construction)

B. Tank Selection

The AST selected for the application must be approved by a fire official, who may require the use of a particular type of tank. Alternatively, if the contents have a high

Table 3 Acceptable Thermal Radiation Separation Distances (ASD) for Aboveground Storage Tanks in Dikes

Dike area (ft^2)[a]	Fire width (ft)	ASD–Buildings 10,000 $BTU/ft^2/hr$	ASD–Persons 450 $BTU/ft^2/hr$
100	10	10	60
200	14	13	80
300	17	15	90
400	20	17	107
500	22	18	115
600	24	21	122
700	26	22	130
800	28	24	140
900	30	25	150
1,000	32	27	160
2,000	45	37	210
3,000	55	44	245
4,000	63	50	275
5,000	71	58	305
6,000	77	60	325
7,000	84	65	350
8,000	89	68	365
9,000	95	73	390
10,000	100	76	400
20,000	141	105	565
30,000	173	125	640
40,000	200	145	720
50,000	224	165	800
60,000	245	175	865
70,000	265	190	905
80,000	283	200	960
90,000	300	215	1020
100,000	316	220	1030

[a]See 24 CFR 51 for dike areas larger than 100,000 ft^2.
Source: Ref. 1.

flash point temperature or enough area is available between the tank and its exposures, any tank acceptable to the jurisdiction may be used. A risk assessment of the tank and its potential impact to the site and community should be included in the decision-making process.

The first criterion is whether the tank was designed and constructed in accordance with nationally recognized standards. This is a requirement of the model fire codes and NFPA standards. Aboveground storage tanks are either shop fabricated or field erected. Shop-fabricated ASTs are constructed to one of the following standards:

Underwriters Laboratories (UL), UL 80 (Steel Tanks for Oil Burner Fuel)
UL 142 (Steel Aboveground Storage Tanks for Flammable and Combustible Liquids)
UL 2080 (Outline of Investigation for Fire Resistant Tanks for Flammable and Combustible Liquids)
UL 2085 (Protected Aboveground Tanks for Flammable and Combustible Liquids)
UL 2244 (Aboveground Flammable Liquid Tank Systems)
UL 2245 (Below-Grade Vaults for Flammable Liquid Storage Tanks)
Southwest Research Institute (SwRI), SwRI Test Procedure 95-03 (Multi-Hazard Rating for Protected Secondary Containment Aboveground Storage Tanks for Flammable and Combustible Liquids)
SwRI Test Procedure 93-01 (Testing Requirements for Protected Aboveground Flammable Liquid/Fuel Storage Tanks)

Part of the acceptance process is ensuring that shop-fabricated tanks are listed by a third-party testing laboratory. When nationally recognized testing laboratories list a tank, it means the manufacturer has demonstrated the ability to produce a product that complies with the requirements of the design standard. It also represents some assurance that the tank has been evaluated for reasonably foreseeable risks associated with the product.

Fire officials should require the use of listed shop-fabricated ASTs. Aboveground storage tanks that are not constructed to UL, SwRI or other equal requirements can present a safety hazard to firefighters and the public. Tanks must be constructed of proper materials and designed with an adequate means of relieving internal pressure due to pool or exposure fires. Incorrectly designed and constructed tanks have caused firefighter deaths. In August 1976, three firefighters were killed when a 6,000-gal aboveground storage tank exploded during a service station fire in Gadsden, Ala. The tank emergency vent was grossly undersized and padlocked shut [3]. Listed shop-fabricated aboveground storage tanks limit the risk of needless tragedies.

The second criterion of tank approval is determining the type of tank that is required for the application. In many cases, the model fire codes and NFPA standards specify acceptable tanks. For example, the 1997 Uniform Fire Code only allows underground storage tanks, aboveground storage tanks in below-grade vaults or protected ASTs at vehicle fueling stations [4]. Design standards also specify the use of certain types of tanks. For example, aboveground storage tanks used as generator-base tanks must meet the required UL 142 top-loading test. This test exceeds the structural-loading test required for conventional steel aboveground storage tanks. The test is performed because generator base tanks are commonly used as generator set foundations. Generator sets are systems designed to provide standby or emergency power for important building features such as emergency lighting in hospitals and high-rise buildings. Diesel engines are commonly used to drive the generator so a base tank is provided. The base tank serves as both the storage tank and as a base for the generator and engine.

Jurisdictions also may establish tank requirements in the ordinance adopting the model fire code. The local fire code amendments may require the use of a particular AST, regardless of its application.

Codes and standards governing tank installation are minimum standards. In certain cases, the code-specified separation distances are derived from anecdotal information such as fire loss history, or actual or interpreted fire test data. The values specified in the codes and standards do not contemplate all particular local conditions. For example, if a community's zoning laws do not address aboveground hazardous-materials storage, the fire official will need to determine if an ASD is provided between the tank and nearby single or multiple-family dwellings.

One method for determining an acceptable separation distance is found in 24 CFR 51. The separation distances and their application are U.S. Department of Housing and Urban Development regulations. The separation distances in this regulation were developed using flammable-liquid pool fire models. The specified thermal-flux values are based on peer-reviewed scientific data for exposure of persons and buildings to varying levels of radiant energy. Application of these distances should provide an adequate separation between the tank and the public in most cases. If these separation distances can be met, a non-insulated steel tank may be acceptable. If they cannot be satisfied, an underground, vaulted, protected or fire-resistant tank should be considered.

The fire official will need to decide if the specified storage tank is acceptable when a model fire code or jurisdiction's adopting ordinance does not require a particular tank type for the given application. Part of the process is to assess the risk of the tank to the safety of firefighters and the community. The risk assessment should include a review of the code or standard addressing tank installations. A tank separated from property lines, buildings, building openings, public ways and egress paths using the requirements in a model code or standard may be acceptable to the jurisdiction. In other cases, the design may not offer the community sufficient safety to limit the consequences of a flammable vapor cloud, fire or explosion.

An AST risk assessment requires a review of several variables. Table 4 identifies variables and key issues when considering a request to install an aboveground storage tank. The variable list is not all-inclusive.

The burden of demonstrating the safety and code compliance of the tank installation belongs to the design professional. The design professional can be a registered professional engineer, licensed architect, or it may be the tank installer. A fire official can require the permit applicant to obtain a technical report prepared

Figure 2 An uninsulated aboveground storage tank may be acceptable to the jurisdiction if it is located far enough from potential exposures.

by a third party. For example, Section 103.1.1 of the 1997 Uniform Fire Code allows the fire chief to obtain technical assistance:

> To determine the acceptability of technologies, processes, products, facilities, materials and uses attending the design, operation or use of a building or premises subject to the inspection of the [fire] department, the chief is authorized to require the owner, or the person in possession or control of the building or premises, to provide, without charge to the jurisdiction, a technical opinion and report. The opinion and report shall be prepared by a qualified engineer, specialist, laboratory, or fire-safety specialty organization acceptable to the chief and the owner and shall analyze the fire-safety properties of the design, operation or use of the building or premises and the facilities and appurtenances situated thereon, to recommend necessary changes. The chief is authorized to require design submittals to bear the stamp of a professional engineer.*

---

*Reproduced from the 1997 edition of the *Uniform Fire Code*™, copyright© 1997, with the permission of the publisher, the International Fire Code Institute. The International Fire Code Institute assumes no responsibility for the accuracy or completion of the summary provided therein.

Table 4 Risk Assessment Variables When Selecting an Aboveground Storage Tank

Variable	Consideration
Content	What are the Øash point and boiling point temperatures, vapor pressure, heat of combustion, and Øammable range of the tank contents? Flammable liquids are easily ignited and large clouds of Øammable vapor can develop that can contact remote ignition sources. Combustible liquids require higher energy ignition sources when compared to Øammable liquids. Combustible liquids have higher heats of combustion when compared to Øammable liquids. Are the tank contents stored at ambient temperatures or heated? If the contents are heated, what controls are provided to prevent the heater from becoming an ignition source? Can the heater raise the liquid above its autoignition temperature? Are the tank contents unstable or reactive materials at normal temperature or if tank is involved in a pool or exposure Ære? Unstable or reactive materials require special consideration such as for emergency venting and tank construction.
Capacity	The duration of a pool Ære depends on the amount of product in the tank. The potential duration for a pool Ære supplied by a leak from a 10,000-gal tank is greater than the same Ære fed from a 3,000-gal tank.
Transfer of tank contents	How is liquid introduced into the tank and transferred to the point of use? Is the liquid transferred using pumps or by gravity dispensing? Is the liquid transferred through pipes or hoses? Hoses have a greater potential for leaks than piped systems.
Containment design	Tanks located within a dike drainage system have a low risk of Ære involvement. Drainage systems are designed to remove and discharge spilled product to a remote containment area. Containment around the tank will not allow a spill to spread. However, if a pool Ære occurs, the tank will be subjected to intense Ære exposure. The area of a Øammable vapor cloud and the thermal Øux of a pool Ære depends on the area of the containment. Containment with a shallow depth and large area will generate more vapors and create a larger pool Ære (and larger thermal Øux). Conversely, a deep containment with a small surface area will generate a smaller volume of vapor and produce thermal fluxes that impact a smaller area, but may burn for a longer duration.
Location of tank openings	Openings located below the tank liquid level are potential sources for leaks. Valves and Ættings are potential leak sources and can malfunction. Are valves manually operated or designed to fail close if involved in Ære? Openings above the liquid line present a lower probability of leakage.

Table 4 Continued

Variable	Consideration
Potential exposures from tank	If a leak occurs, what is the potential leakage rate and how can this affect surrounding exposures? If the tank were involved in a pool Ære, how would this impact adjacent equipment or operations?
Location of tank	What is the location of the tank in relation to buildings, building openings, property lines and public ways? Is the tank located in an area that is densely populated or congested with other buildings or structures? If the emergency vent is disabled, who and what will be impacted if the tank explodes?
Operations and maintenance	How are the tank, transfer pumps, pipes, hoses and dispensers maintained? Is maintenance a scheduled activity or performed only when repairs are required? What is the level of training and competency of persons operating the equipment? What is the frequency of inspecting the tank and equipment?
Capabilities of ÆreÆghting resources	Does the Fire Department have sufÆcient resources to safely apply water or Ære Æghting foam on the tank and the containment area? Are ÆreÆghters trained to handle petroleum Ære incidents and what is their training frequency? Is the tank protected by automatic or manual water or foam-water spray systems?

This report is a technical analysis of the proposed design to determine its adequacy and compliance with a community's adopted fire code. The report should contain a risk assessment of the proposed design and the engineering and administrative controls provided to reduce the frequency—and limit the consequences—of a leak, pool fire and tank explosion. The cost for preparing the technical report is the responsibility of the permit applicant. The fire official is responsible for accepting or rejecting the report findings.

C. Access and Water Supply

When a tank is located in an area where buildings are not present, such as a remote construction site or quarry, a means must be provided so firefighters can safely gain access to the site. The site must also have an adequate water supply so firefighters can protect the tank or the exposures adjacent to the tank.

The model fire codes have requirements for fire department access to a facility. While the requirements vary in each code, they all generally provide for:

All-weather access roadways designed to support the imposed load of firefighting apparatus

Sufficient vertical clearance to enable fire trucks and other needed equipment access onto the site

Figure 3 An adequate, reliable water supply is required for firefighters to successfully control and extinguish a fire involving the contents of an aboveground storage tank.

Roadways constructed with adequate width, grades, and turning radii so firefighting apparatus can maneuver around the site and tank
A method allowing the fire department access if gates or other impediments secure the site

A firefighting water supply is also required. If an AST is installed on a building site constructed in a community that has adopted a fire code, the site will most likely have a water supply source for the building. This could include fire hydrants connected to a water main or a water tank intended for firefighting purposes.

Neither the National Fire Protection Association (NFPA), the American Petroleum Institute (API), the American National Standards Institute (ANSI), nor the American Society for Testing and Materials (ASTM) publish a standard for determining the fire flow—the flow rate of a water supply that is available for firefighting—for aboveground storage tank installations. Therefore, the calculation of a required fire flow for a tank installation should be based on either cooling the tank or cooling the tank and extinguishing the pool fire.

The required fire flow will be influenced by this decision. If an extinguished fire is the goal, then a larger fire flow will probably be required. The required fire flow will also be influenced by the material in the storage tank. For example, esters and ketones present a greater fire control and suppression challenge than simple hydrocarbons. These materials exhibit higher heats of combustion. Esters and ketones are also polar solvents. Polar solvents exhibit good solubility with water. Therefore, firefighting foams that are compatible with these materials are normally required.

Another consideration is the proximity of other tanks or piping. The protection of the primary tank and the exposure tanks requires an analysis of:

The pool fire or jet fire (a high-velocity flame that is directed to a particular target)
The arrangement of the drainage or containment system
The proximity of the exposure tanks to the primary tank
The hazard characteristics of the materials stored

This type of analysis is beyond the scope of this chapter. Fire officials should seek technical assistance in such cases.

Water is a very effective firefighting agent. Water has a latent heat of vaporization (i.e., converting water into steam) of 970 BTU/gal. Properly applied, water reduces both the surface and internal temperatures of a fire-involved tank. Lowering the tank temperature prevents the development of excessive internal pressures and cools the steel so it will not rapidly lose its structural integrity. If elevated, water applied to the structural supports protects the tank from possible collapse. Water will dilute water-miscible liquids (e.g., alcohols). Diluting a water-miscible liquid reduces its heat of combustion and will eventually extinguish the fire.

A tank filled with liquid will absorb heat from a pool or exposure fire. The time required for the contents to absorb heat at any given rate principally depends upon the diameter of the tank. A non-insulated tank involved in a pool fire can be expected to absorb heat at a rate of at least 20,000 BTU/ft^2/hr of exposed surface wetted by its contents [5]. Assuming the liquid inside the tank is stored at a temperature of 70°F, and using this heat-absorption rate, volatile liquids in a 15-ft-diameter tank with a volume of 20,000 gal are heated to a temperature of 100°F in about 10 min.

Applying water spray to a vessel involved in fire will reduce the heat-input rate to a value of 6000 BTU/ft^2/hr of exposed surface wetted by the contents of the tank. This reduced heat-input rate is based on a water application rate of 0.20 gal/min/ft^2 (gpm/ft^2). At this heat absorption rate, volatile liquids in a 15-ft-diameter tank with a volume of 20,000 gal are heated to a temperature of 100° F in about 34 min [6].

Water should be applied at a minimum rate of 0.25 to 0.50 gpm/ft^2. The required density depends on the orientation of the tank, the proximity of exposures, and if a pool fire or jet fire are contemplated. For a 1000-ft^2 area, this equals an application rate of 250 to 500 gpm. The duration of the water supply should consider the properties and quantities of the materials and the anticipated effect of the firefighting resources. A supply capable of delivering water for several hours may be required [7].

To achieve quick fire suppression, improve firefighter safety, and limit the demand and duration of the water supply, the fire flow can be calculated assuming the use of firefighting foam. Low-expansion firefighting foam is used principally to extinguish burning flammable or combustible liquid spills or tank fires. Its application allows firefighters to extinguish these fires progressively. A foam blanket creates a vapor-sealing barrier on the liquid surface, preventing the development and spread of flammable vapors. Foam cools and extinguishes pool fires by vaporizing water in the solution.

Different types of firefighting foam agents are available. Foam concentrates can be formulated for specific types of flammable and combustible liquids, or for all types of liquid fires. The selection of the appropriate foam concentrate should be decided using information provided by the chemical manufacturer or listed in the chemical's material safety data sheet. Foam concentrate is mixed in solution with water at a certain concentration. Foam concentrates for flammable and combustible liquids are typically 3 percent or 6 percent solutions. The amount of foam concentrate required for firefighting operations must be calculated as part of the fire flow.

NFPA 11 (Standard for Low-Expansion Foam) contains requirements for determining the minimum fire flow when using firefighting foam. The requirements cited in this chapter assume the use of attack hand lines on fixed-roof tanks. Foam attack hand lines are at least 1 ½-in.-diameter firefighting hose with nozzles that are used by firefighters to direct the foam-water solution onto the liquid pool. Foam attack hand lines are an acceptable protection method for fixed-roof tanks less than 30 ft in diameter or less than 20 ft in height [8]. Foam monitors are required for tanks over 20 ft in height. Foam monitors are commonly large-volume water nozzles that are connected to a piping system. These monitors are capable of delivering several thousand gallons of foam-water solution per minute.

The fire flow for an aboveground storage tank is based on the exposed surface area of the tank and the area of the secondary containment or dike. The equations for calculating the surface area of rectangular and cylindrical tanks are:

Rectangular tanks:

$$SA_{rectangular} = 2[(lw) + (lh) + wh)]$$

where $SA_{rectangular}$ is the surface area of a rectangular, horizontal tank (ft^2); "*l*," "*w*," and "*h*" are the length, width, and height of the tank (ft).

Cylindrical tanks:

$$SA_{cylindrical} = 2\,\pi r(r + h)$$

where $SA_{cylindrical}$ is the surface area of a vertical or horizontal tank shaped like a cylinder (ft^2); π is 3.14; "r" is the radius of the tank (ft), and "h" is the height or length of the tank (ft). The following examples illustrate the method of calculating the surface area of horizontal and vertical tanks.

Example: Calculate the surface area of 1500-gal rectangular aboveground storage tank containing unleaded gasoline. The tank is located within secondary containment that measures 10 ft wide by 40 ft in length. The tank is elevated 2 inches. The tank and containment dimensions are shown in Figure 4.

$$l = 11 \text{ ft}$$
$$w = 5 \text{ ft}$$
$$h = 4 \text{ ft}$$
$$SA_{rectangular} = 2[(lw) + (lh) + wh)]$$
$$SA_{rectangular} = 2[(11 \text{ ft} \times 5 \text{ ft}) + (11 \text{ ft} \times 4 \text{ ft}) + (5 \text{ ft} \times 4 \text{ ft})]$$
$$SA_{rectangular} = 238 \text{ ft}^2$$

The area of the containment is calculated by multiplying the length by the width:

$$\text{Containment area} = (l)(w)$$
$$\text{Containment area} = (40 \text{ ft}) \times (10 \text{ ft})$$
$$\text{Containment area} = 400 \text{ ft}^2$$

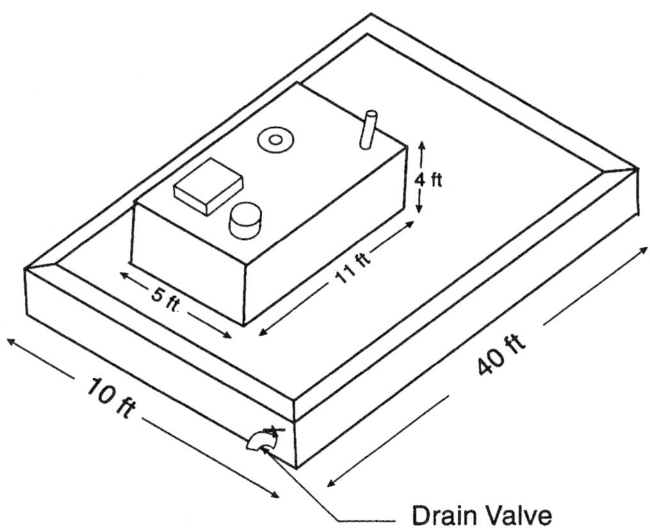

Figure 4 1500-gal rectangular tank.

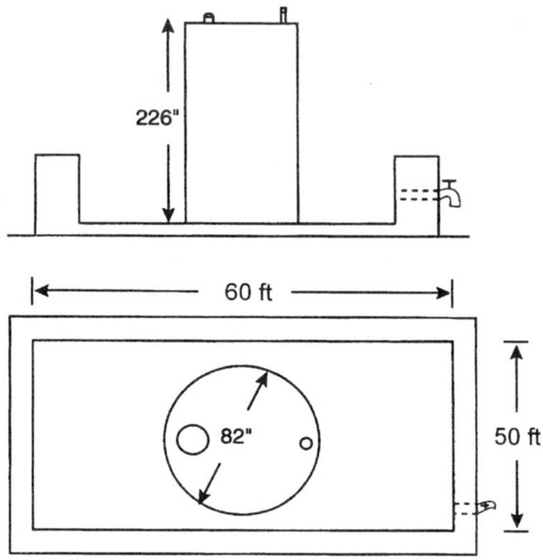

Figure 5 5000-gal vertical tank.

Example: Calculate the surface area of a 5000-gal vertical storage tank storing No. 2 diesel fuel. The tank is elevated 2 inches. The tank is located in secondary containment that measures 60 ft wide by 50 ft long. The tank and containment dimensions are illustrated in Figure 5.

$$h = 226 \text{ in.} \times \frac{1 \text{ ft.}}{12 \text{ in.}} = 18.83 \text{ ft.}$$

r = ^{1}§ the diameter of the tank, or 82 in $\times \frac{1 \text{ ft.}}{12 \text{ in.}} \times 0.5 = 3.42$ ft radius

$$SA_{cylindrical} = 2\pi r(r + h)$$
$$SA_{cylindrical} = 2 \times 3.14 \times 3.42 \text{ ft}(3.42 \text{ ft} + 18.83 \text{ ft})$$
$$SA_{cylindrical} = 21.48 \text{ ft} \times (22.25 \text{ ft})$$
$$SA_{cylindrical} = 478 \text{ ft}^2$$

The area of the containment is calculated by multiplying the length by the width:

Containment area = $(l)(w)$

Containment area = (60 ft) × (50 ft)

Containment area = 3000 ft^2

Requirements for the NFPA 11 minimum application rate and discharge time are in Table 5. These application rates and the minimum discharge duration are

Table 5 Foam Handline and Monitor Protection for Fixed-Roof Storage Tanks Containing Hydrocarbons

Hydrocarbon type	Minimum application rate (gPM/ft²)	Minimum discharge time (min)
Flash point temperature between 100°F and 140°F	0.16	50
Flash point temperature below 100°F or liquids heated above their Øash point	0.16	65
Crude petroleum	0.16	65

Note 1. Included in this table is gasohol and unleaded gasoline solution containing not more than 10 percent oxygenated additive by volume. When the oxygenated additive content exceeds 10 percent by volume, protection is normally in accordance with 3-2.2.4. Certain nonalcohol resistant foams might be suitable for use with fuels containing oxygenated additives of more than 10 percent by volume. The foam manufacturer should be consulted for speciÆc listings or approvals.
Note 2. Flammable liquids having a boiling point less than 100°F might require higher rates of application. Suitable application rates should be determined by test. Flammable liquid with a wide range of boiling points might develop a heat layer after prolonged burning and then can require application rates of 0.20 GPM/ft^2 or more.
Note 3. Care should be taken in applying portable foam streams to high-viscosity materials heated above 200°F. Good judgment should be used in applying foam to tanks containing hot oils, burning asphalt, or burning liquids that have a boiling point temperature above the boiling point of water. Although the comparatively low water content of foams can beneÆcially cool such fuels at a slow rate, it can also cause violent frothing and "slop over" of the tank contents.*
Source: Ref. 8.

based on extrapolations of test experiences and the listings issued by the nationally recognized testing laboratories for foam concentrates and foam-producing equipment.

Using the sum of the surface area of the tank and containment, the appropriate discharge density and duration in Table 5 are applied to determine the required fire flow and duration. Multiply the aggregate area by the required density to determine the fire flow. Next, the fire flow is multiplied by the percent volume of the foam solution and the discharge duration to determine the amount of foam/water solution required for firefighting. The following examples illustrate the calculations.

Example: A 1500-gal rectangular aboveground storage tank storing unleaded gasoline has a surface area of 238 ft^2 and a containment area of 400 ft^2. Calculate

---

the fire flow and duration. The aggregate exposed area of the tank and containment is:

$$238\ ft^2 + 400\ ft^2 = 638\ ft^2$$

The tank stores unleaded gasoline without oxygenated additives. The material safety data sheet states the flash point temperature is lower than 100°F. The requirements in Table 5 specify a minimum discharge density of 0.16 gpm/ft^2 for a minimum discharge duration of 65 min.

$$638\ ft^2 \times 0.16\ gpm/ft^2 = 102\ gpm \text{ of foam/water solution}$$

For a 65-min period, a minimum of 6635 gal of foam/water solution is required for firefighting purposes. If 3 percent foam concentrate is used, a minimum of 199 gal of foam concentrate is required.

Example: A 5000-gal vertical storage tank storing No. 2 diesel fuel has an exposed surface area of 478 ft^2 and a secondary containment area of 3000 ft^2. Calculate the fire flow and duration. The aggregate area of the tank and containment is:

$$478\ ft^2 + 3000\ ft^2 = 3478\ ft.^2$$

The material safety data sheet states No. 2 diesel fuel has a flash point temperature of 123°F. Based on the liquid flash point temperature, the minimum discharge density is 0.16 gpm/ft^2 for a minimum discharge duration of 50 min.

$$3478\ ft^2 \times 0.16\ gpm/ft^2 = 556\ gpm \text{ of foam/water solution}$$

For a 50-min period, at least 27,824 gal of foam/water solution is required for firefighting purposes. Assuming the use of 3 percent foam, a minimum supply of 835 gal of concentrate is required.

D. Tank Foundations

Tank foundations should be designed to evenly support a tank filled with product and the equipment attached to it (Fig. 6). Improperly designed foundations can cause uneven stresses on tanks and equipment, causing leaks or complete failures. The design of the foundation is influenced by several factors. They include the tank weight, the specific gravity of the stored liquid, the tank's orientation (horizontal or vertical), the load-bearing capacity of the soil and drainage around the tank. Foundation design should include an analysis of the site drainage, the presence of flood plains and an analysis of the soil conditions. In certain areas, an analysis of the foundation under seismic loading may be required. Because of these variables, a registered professional engineer should perform the design of the foundation.

The primary criterion for tank foundation design is the load-bearing capacity of the soil. This ranges from 1000 lb/ft^2 (psf) in soft clay, to over 4000 psf for coarse, compacted gravel. A 12,000 gal vertical tank, 8 ft in diameter, which contains gasoline (specific gravity of 0.75), exerts a force of about 2000 psf on its foundation. The unit load on the foundation from a horizontal tank depends on the number of saddles and the arrangement of the footings [9].

A number of manufacturers are constructing precast concrete slabs as part of the tank installation. These slabs are constructed to meet a 28-day, 2500 to 4000 psi compressive-strength requirement, depending on the tank's construction and size. The slabs are engineered to meet the model building code requirements for snow, wind and seismic loads.

If tanks are elevated more than 12 in. above grade, they will require additional protection. The structural supports should have a minimum two-hour fire resistance or should be protected with a water spray system. When a fire-resistant assembly or coating is applied on the columns, it should meet the requirements of ASTM E-1529 (Standard Test Method for Determining Effects of Large Hydrocarbon Pool Fires on Structural Members and Assemblies). Fire-resistant coatings

Figure 6 A tank foundation is critical for maintaining the design life of the tank. This foundation is designed to support the loads imposed by a 12,000-gal protected aboveground storage tank.

or systems tested to this standard are capable of withstanding the high temperatures and thermal fluxes that are characteristic of pool fires.

E. Secondary Containment

Aboveground storage tanks require a means to impound spills resulting from failures or leaks from tanks, pipes, equipment, or overfills. The model fire codes specify secondary containment for almost every type of outdoor aboveground tank installation. Federal laws also require the installation of secondary containment. Under the Oil Pollution Act, 40 CFR Part 112, aboveground storage tanks larger than 660 gal. that are located near navigable waterways require secondary containment. The regulation also requires the facility management to prepare a Spill Prevention Control and Countermeasures (SPCC) plan documenting the administrative and engineering controls provided to prevent or respond to a spill. The U.S. Environmental Protection Agency (EPA) regulates tanks storing flammable liquids at hazardous waste treatment, storage and disposal facilities. Such tanks require secondary containment, leak detection and other safety controls pursuant to 40 CFR 264.193.

Primary containment consists of the inside portion of a container that comes into immediate contact on its inner surface with the material being stored. A single-wall tank is a method of primary containment. Secondary containment is that level of containment that is external to and separate from the primary containment [10]. The requirement for secondary containment can be satisfied using a number of methods. Secondary containment can be constructed as a liquid-tight barrier or "dike" around the tank. Another option is to construct the tank with integral secondary containment, i.e., a "tank inside of a tank."

Incidents of petroleum product spills reported to the United States Coast Guard are shown in Table 6. Data for the period from 1988 through 1992 indicates that tanks themselves are not a significant contributor to the overall number of spills [11]. The number of spills caused by tank and equipment failures is low when compared to unknown causes, which increased each year of the reporting period. This is most likely attributed to human errors during tank filling, operations and maintenance. The data shows that significant and possibly catastrophic failures occur annually. Tens of thousands of gallons of flammable and combustible liquids are spilled each year. Given this data, the fire codes require secondary containment around tanks and process equipment.

Containments are commonly designed as an integral element of tank foundations. The designer must evaluate the material used to construct or line the containment. These materials must be liquid-tight and resistant to the stored chemicals. Containments must be designed to resist buoyant forces when located in areas with high water tables. Therefore, a registered professional engineer should su-

Table 6 Reported Spill Incidents Caused by Tank/Container and Equipment Failure or Unknown Causes–1988 Through 1992

Year	Cause	% of spill incidents	No. of spills	Average spill size (gal)	Median spill size (gal)	Maximum spill size (gal)
1992	Container/tank failure	0.5	46	197	13	6,500
	Equipment failure	2.9	279	238	5	12,000
	Unknown	91.4	8672	194	2	145,025
1991	Container/tank failure	2.6	223	236	8	9,000
	Equipment failure	17.4	1490	159	2	28,056
	Unknown	49.1	4205	180	2	500,000
1990	Container/tank failure	2.6	211	333	15	36,657
	Equipment failure	13.1	1071	280	5	84,000
	Unknown	48.8	3993	1127	2	3,900,000
1989	Container/tank failure	3.4	228	3217	10	292,000
	Equipment failure	13.0	862	175	5	16,800
	Unknown	44.2	2926	99	4	43,000
1988	Container/tank failure	3.5	174	285	20	7,200
	Equipment failure	15.4	770	331	10	168,000
	Unknown	43.4	2168	1350	5	2,041,662

Source: Ref. 11.

pervise the design of the secondary containment if it is not listed as a secondary containment tank. A properly designed and maintained secondary containment should:

Have sufficient volume to contain a release from the largest tank completely filled with product and, when required, sufficient freeboard to accommodate the rainfall from a 24-hr/25-yr storm.

Have sufficient strength to withstand the liquid hydrostatic pressure on the containment floor and walls. The containment should be constructed on a foundation or base capable of providing support to the secondary containment system, resisting pressure gradients above and below the system, and preventing failure due to settlement, compression or uplift.

Be constructed using materials that are compatible with the stored chemicals to form a liquid-tight containment. Containments should be constructed of or lined with materials that are compatible with the liquid to be placed in the tank. The secondary containment also must have sufficient strength and thickness to prevent failure from pressure gradients (e.g., static head and external hydrological forces), physical contact with the liquid to which it is exposed, and climatic conditions.

Provide a means to remove rainfall or prevent its entry into the containment.
Provide sufficient access for maintenance and firefighting activities.

Experience has shown that in most cases of tank rupture only a portion of the liquid in the tank is lost and, in many cases, the tank retains most of the liquid. When evaluating a containment design, consider that a leak in the side of the tank may form a horizontal fluid jet and may in fact jump the containment if the tank is too close. This effect is known as a spigot flow. To avoid this, the containment wall should be far enough from the tank to prevent a fluid jet from jumping over or, alternatively, the containment should be surrounded by walls sloped in toward the containment foundation.

The geometry of the tank and containment dictates how the volume is calculated. For rectangular and square tanks and containments the formula for liquid volume (V) is:

$$V = (l)(w)(h)7.48\frac{\text{gal}}{\text{ft}^3}$$

where l = length (ft), w = width (ft), and h = height (ft).

The formula for circular tanks and containments is:

$$V = \frac{\pi}{4}D^2H7.48\frac{\text{gal}}{\text{ft}^3}$$

where $\pi = 3.14$, D^2 = the square of the tank diameter (ft), and H = height or length of the circular tank.

Depending on the jurisdiction's fire code and environmental regulations, the containment may also be required to hold a certain amount of rainwater. Some regulations require that the containment design include a sufficient freeboard to hold rainwater from a 24-hr/25-yr storm plus the contents of the largest tank. Data on this rainfall frequency is contained in a May, 1961 publication titled "Technical Paper No. 40—Rainfall Frequency Atlas of the United States." Prepared by the U.S. Department of Commerce, the publication is no longer in print. However, the information is available from most regional offices of the National Weather Service.

Figure 7 illustrates an 8000-gal horizontal aboveground storage tank located inside a rectangular secondary containment. Calculate the volume of the containment and determine if it has sufficient freeboard for a 1-ft rainfall from a 24-hr/25-yr storm.

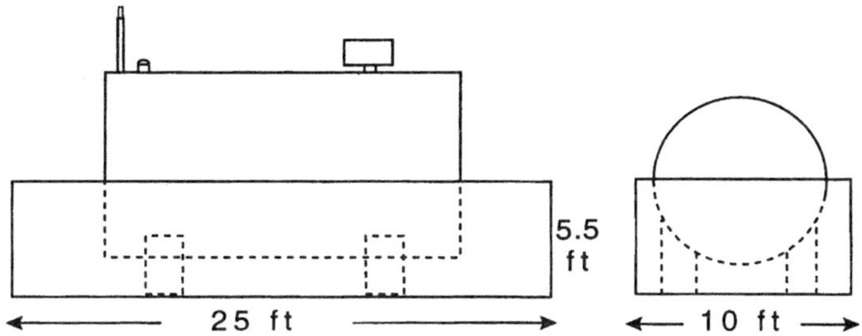

Figure 7 8000-gal horizontal aboveground storage tank located inside of a rectangular secondary containment.

The volume of the containment is:

$$V = (l)(w)(h)7.48\frac{\text{gal}}{\text{ft}^3}$$

$$V = (25\ \text{ft})(10\ \text{ft})(5.5\ \text{ft})7.48\frac{\text{gal}}{\text{ft}^3}$$

$$V = (1375\ \text{ft}^3)7.48\frac{\text{gal}}{\text{ft}^3}$$

$$\text{V} = 10{,}285\ \text{gal}.$$

The design will adequately contain the volume of the tank. To determine if the containment can accommodate the rainfall from a 24-hr/25-yr storm, calculate the volume of a 1-ft rainfall and deduct this amount from the containment volume. If the containment volume is greater than the tank volume, the design will contain the contents of the largest tank plus a 24-hr/25-yr storm.

$$V_{\text{rain}} = (l)(w) \times \text{rainfall amount (ft)} \times 7.48\frac{\text{gal}}{\text{ft}^3}$$

$$V_{\text{rain}} = (25\ \text{ft})(10\ \text{ft})(1\ \text{ft}) \times 7.48\frac{\text{gal}}{\text{ft}^3}$$

$$\text{V}_{\text{rain}} = 1{,}870\ \text{gal}.$$

Volume of containment:	10,285 gal
Volume of rain:	−1,870 gal
	8,415 gal

The containment will hold the tank contents and the rainfall from a 24-hr/25-yr storm.

The available capacity of a secondary containment is reduced when two or more tanks are located inside it. The secondary containment or dike system must include a volume sufficient to offset the area consumed by the tanks. NFPA 30 (Flammable and Combustible Liquids Code), Section 2-3.4.3 (b) requires:

> The volumetric capacity of the diked area shall not be less than the greatest amount of liquid that can be released from the largest tank with the diked area, assuming a full tank. To allow for volume occupied by tanks, the capacity of the diked area enclosing more than one tank shall be calculated after deducting the volume of the tanks, other than the largest tank, below the height of the dike.*

To compensate for the volume af liquid displaced by the tanks, subtract the amount of unavailable volume from the available volume. Table 7 contains depth factors for cylindrical, horizontal tanks (Fig. 8). The depth factor can be used to calculate the number of gallons displaced by the tank inside the containment. The liquid height below the containment wall, the diameter of the tank, and the containment wall height are required.

$$\text{Depth factor} = \frac{H}{D}$$

Figure 9 illustrates a 12,000-gal and 10,000-gal horizontal AST inside secondary containment. The tank diameter of the 10,000 gal AST is 8 ft, and 3 ft of the tank diameter is below the containment wall. The height of the containment walls is 5 ft. The 10,000-gal AST is used when calculating the volume displaced because it is the smallest tank inside of the secondary containment. Calculate the volume of the containment displaced by the 10,000-gal tank.

$$\text{Depth ratio} = \frac{H}{D}$$

$$\text{Depth ratio} = \frac{3\text{ ft}}{8\text{ ft}}$$

$$\text{Depth ratio} = 0.38$$

Table 7 indicates a liquid depth-tank diameter factor of 0.3487 for a ratio of 0.38. The tank volume is 10,000 gal.

---

*Reprinted with permission from NFPA 30, *Flammable and Combustible Liquids Code*, Copyright© 1996, National Fire Protection Association, Quincy, MA 02269. This reprinted material is not the complete and official position of the National Fire Protection Association, on the referenced subject which is represented only by the standard in its entirety.

Table 7 Horizontal Tank Ullage by Tank Diameter and Liquid Depth Percentage

Ratio	Depth factor	Ratio	Depth factor	Ratio	Depth factor	Ratio	Depth factor	Ratio	Depth factor
0.02	0.0048	0.22	0.1631	0.42	0.3986	0.62	0.6513	0.82	0.8776
0.04	0.0134	0.24	0.1845	0.44	0.4238	0.64	0.6579	0.84	0.8967
0.06	0.2450	0.26	0.2066	0.46	0.4491	0.66	0.7002	0.86	0.9149
0.08	0.0375	0.28	0.2292	0.48	0.4745	0.68	0.7241	0.88	0.9320
0.10	0.0520	0.30	0.2523	0.50	0.5000	0.70	0.7477	0.90	0.9480
0.12	0.0680	0.32	0.2759	0.52	0.5255	0.72	0.7708	0.92	0.9625
0.14	0.0851	0.34	0.2998	0.54	0.5509	0.74	0.7934	0.94	0.9755
0.16	0.1033	0.36	0.3241	0.56	0.5762	0.76	0.8154	0.96	0.9865
0.18	0.1224	0.38	0.3487	0.58	0.6014	0.78	0.8369	0.98	0.9952
0.20	0.1424	0.40	0.3735	0.60	0.6265	0.80	0.8576	1.00	1.000

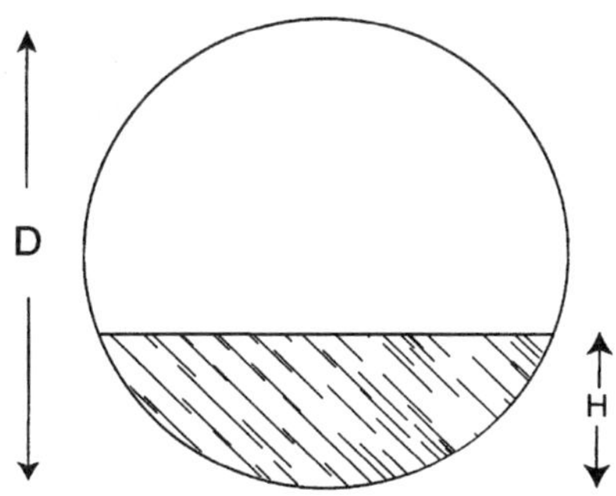

Figure 8 Horizontal tank ullage figure.

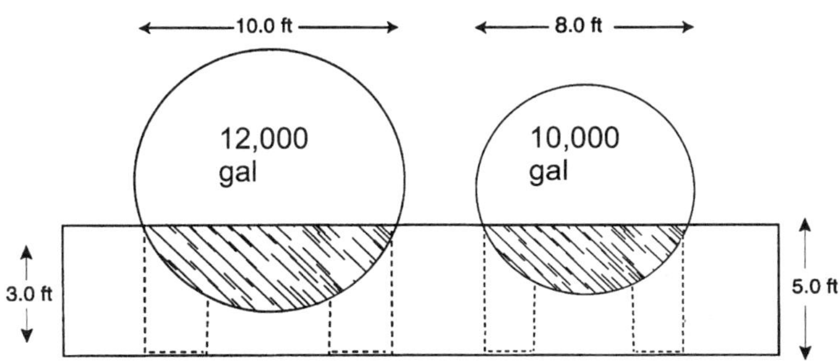

Figure 9 12,000-gal and 10,000-gal horizontal aboveground storage tanks inside of secondary containment.

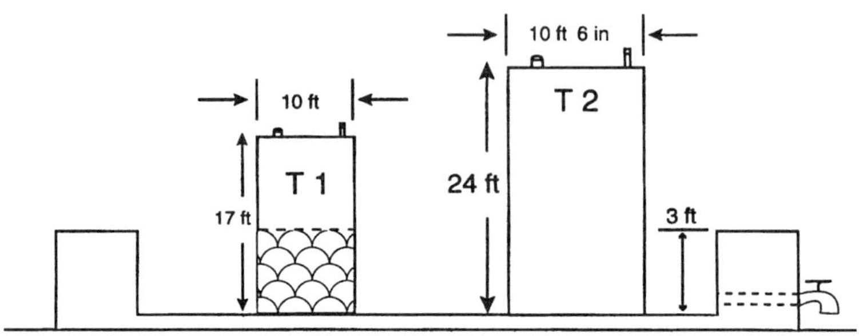

Figure 10 Two vertical aboveground storage tanks.

(10,000)(0.3487) = 3487 gal of containment volume displaced by the tank

The volume displaced by vertical storage tanks is calculated in a similar manner. Figure 10 illustrates two vertical aboveground storage tanks. The volume of tank No. 1 is 10,000 gal. The volume of tank No. 2 is 15,000 gal. The containment area is 1000 ft^2. The containment wall height is 3 ft. The volume displaced by the smaller tank is calculated using the formula for cylindrical tanks. In this example, the height value (H) will be the height of the containment wall. The volume of the containment is:

$$(1000\ \text{ft}^2)(3\ \text{ft})7.48\frac{\text{gal}}{\text{ft}^3} = 22{,}440\ \text{gal}$$

The volume displaced by tank No. 1 is calculated using the following equation:

$$V = \frac{\pi}{4}D^2H7.48\frac{\text{gal}}{\text{ft}^3}$$

$$V = \frac{\pi}{4}(10\ \text{ft})^2(3\ \text{ft})7.48\frac{\text{gal}}{\text{ft}^3}$$

$$V = (235.6\ \text{ft}^3)7.48\frac{\text{gal}}{\text{ft}^3}$$

$$\text{V} = 1762\ \text{gal}$$

Tank No. 1 displaced volume is subtracted from the volume of the containment. If the resulting value is larger than the volume of the largest tank, the secondary containment is adequate:

22,440 gal - 1762 gal = 20,678 gal

The volume of tank No. 2 is 15,000 gal, so the containment can hold the contents of the largest tank.

Containments must be liquid-tight. A commonly asked question is, "What constitutes liquid-tight construction?" One answer is the U.S. EPA requirement for liquid tightness of chemical-resistant liners. 40 CFR 264.221 has requirements for the construction of surface impoundments at hazardous-waste treatment, storage, and disposal facilities. The regulation specifies that containments or chemical liners must have a permeation rate of no greater than 1×10^{-7} centimeter/seconds using a minimum base of 3 ft. of compacted soil. This permeation rate over the 3-ft depth value is about 72 hr. In other words, the containment should be designed to hold the contents for 3 days without any leakage or contamination of the surroundings.

Containments are commonly constructed by erecting a concrete form around the area where the tanks will be located. The concrete form is designed so the walls and foundation become a single, complete containment. This is known as monolithic containment because it prevents joints or seams from becoming leak paths (Fig. 11). Containments can also be constructed using multiple forms. If two or more forms are used, the review should evaluate the water-stopping joints used at the wall and foundation connections. Water-stopping joints should be compatible with the chemicals stored in the tanks. Water-stopping joints should be installed in accordance with the manufacturer's instructions.

Figure 11 A single concrete form was constructed to create a liquid-tight monolithic containment at the connection of the walls and foundation.

One design that should be scrutinized is secondary containment constructed from concrete masonry units, or blocks. This form of containment may lack liquid-tightness because:

Settling separates the blocks and, in some cases, causes cracks that destroy the integrity of the containment wall.
Spalling of the mortar will allow liquid to penetrate through the joints.
It is difficult to construct a liquid-tight wall from concrete blocks because of their porosity and the vulnerability of mortared joints [12].

If rain or snow enters a containment, a means must be provided to remove the water. Installing a valve outside the containment is one means for removing water. NFPA 30 and the model fire codes have requirements for containment-drainage valves. The valve should be located at the lowest elevation of the containment so the water can drain. The valve also must be installed so it can be locked or plugged. This prevents the accidental or malicious release of water or spilled contents from the secondary containment. Unfortunately, occurrences where the drain valve was left open are still reported. Facility management must ensure that the containment drain valve is closed after use (Fig. 12).

To avoid the expense of building a secondary containment or dike area, listed tanks constructed with secondary containment are available. These tanks come in the form of a primary tank enclosed by an outer secondary tank. They can also be constructed as a single-wall tank within a steel containment basin. One form of secondary containment for an aboveground storage tank consists of a steel, single-wall atmospheric tank contained within a secondary containment shell. The primary tank and secondary tank form an annular (or interstitial) space. A means for monitoring leakage into the interstitial space must be provided at the exterior or interior tank walls. The secondary containment shell is constructed of steel on a noninsulated tank [13].

Secondary containment for protected and fire-resistant tanks used for motor-vehicle fueling may be constructed of steel, concrete or other approved materials, depending on how the primary tank is insulated. The containment materials' permeation rate cannot exceed 4×10^{-8} in./sec with the stored fluids. The containment cannot interfere with emergency venting of the interstitial space. To reduce the potential for leaks, all openings that penetrate the secondary containment structure must terminate above the maximum liquid level of the tank [14].

An open or closed steel dike can also provide secondary containment. Closed-top dikes are designed to keep precipitation or debris from entering the dike area. The design of closed-top dikes must not interfere with the operation of the emergency vent. UL 142 specifies that an open-top diked tank requires a dike capacity capable of holding 110 percent of the primary tank's contents. The 110 percent dike-volume requirement may not be adequate if it also is expected to contain rainfall from a 24-hr/25-yr storm. It also may not satisfy certain environmen-

Figure 12 A method of removing water from a secondary containment is required. In this case a valve is used. Note that valve is not locked closed.

tal regulations. For example, some states require a containment to have an available volume of 150 percent of the tank contents.

F. Aboveground Storage Tank Siting Requirements

NFPA 30, NFPA 30A (Automotive and Marine Service Station Code), and the model fire codes have requirements for locating aboveground storage tanks on property. The liquid classification, the type and volume of the tank, and the application can influence where a tank will be sited. While the requirements for types of tanks and controls vary between the different codes and standards, the specified separation distances deviate very little between each code.

Compared to the requirements for bulk storage facilities, fuel supplies for standby or emergency power, and industrial chemical storage, mandates targeting

ASTs that enable motor-vehicle fuel dispensing call for the most restrictive siting distances. Tanks used in industrial applications are usually located in areas that are not easily accessible to the public. Industrial buildings are normally in an area of the community with few or no single- or multiple-family dwellings. Tanks used for industrial processes can be subjected to greater degrees of regulation such as the U.S. Occupational Safety and Health Administration (OSHA) process-safety management requirements. The separation distances for motor-vehicle fueling are greater to create a safety zone between a hazardous activity and the public. Dispensing flammable and combustible liquids into vehicles introduces risks that are not common to other tank uses:

Tanks and dispensers may be accessible to the public. Site security can vary by location.
The tank can be located in an urban setting where the potential impact of a fire or explosion to persons and buildings is greater.
The ability of the user to safely dispense fuel will vary.

Table 8 contains some basic tank-siting requirements for aboveground storage tanks located at service stations accessible to the public. The separation distances cited assume the installation of a 12,000-gal storage tank. The table indicates that noninsulated tanks require larger separation distances when compared to the other tank types. Depending on the adopted fire code, a noninsulated tank may not be permitted. Conversely, fire-resistant tanks, protected tanks and vaulted tanks have more liberal separation distances. Fire-resistant and protected tanks offer two-hour fire resistance from exposure and pool fires.

Table 8 Separation Distances for a 12,000-Gal AST Installed at Public Accessible Service Stations

Tank type	Separation distances
Uninsulated	100 ft from property lines and 50 ft from important buildings on the same property[a] *or* Not permitted at retail service stations accessible to the general public[b]
Protected or Ære resistant	50 ft from property lines 25 ft from important buildings on the same property
Tanks in below-grade vaults	50 ft from property lines 25 ft from important buildings on the same property

[a]NFPA 30A allows insulated ASTs when the separation distances are satisÆed.
[b]BOCA Fire Prevention Code, Standard Fire Prevention Code and the Uniform Fire Code do not allow the use of uninsulated aboveground storage tanks at service stations accessible to the public.

Tanks in vaults may be installed at above- or below-grade, although some codes specifically recognize only below-grade installations. In a below-grade installation, the vault itself is buried underground, and a tank designed for aboveground installation, such as a UL 142 tank is installed. Note that tanks in vaults are not required to be fire-resistant or protected tanks because enclosure of a tank within a vault isolates the tank from external fire exposures and vice versa. Underground vaults are required to be liquid tight to prevent groundwater from seeping into a vault and to prevent any spill that might occur inside of a vault from leaking to soil. Pre-manufactured vaults listed in accordance with UL 2245 are evaluated by UL to verify their secondary containment capabilities. The U.S. EPA has ruled that tanks installed in below-grade vaults can be treated as aboveground storage tanks for the purposes of applying federal UST regulations for required monitoring systems, corrosion protection and similar controls. Underground piping from the vault to the dispenser or process is normally not exempted. In certain states, tanks in vaults are exempt from taxation that supports leaking UST cleanup tax funds.

G. Tank Vents

Listed aboveground storage tanks are manufactured with openings for both normal and emergency vents. These vents protect the tank from implosion or explosion due to product movement, temperature changes and fire exposure. The model fire codes and NFPA 30 contain detailed design, operation and maintenance requirements for normal and emergency vents. Many of the model fire code requirements are based on API Standard 2000 (Venting Atmospheric and Low-Pressure Storage Tanks-Nonrefrigerated and Refrigerated). The standard covers the normal and emergency venting requirements for aboveground liquid petroleum or petroleum product storage tanks, and aboveground and underground refrigerated storage tanks designed for pressures from vacuum through 15 psig [15].

1. Normal Vents

Storing liquid inside a tank creates pressure on the tank shell and heads. Adding liquid increases the pressure inside of the tank. When liquid is removed, the pressure inside of the tank is lowered. To regulate internal pressure and enable filling and emptying, a tank must be adequately vented. A normal vent is installed to prevent excessively positive or negative pressure on the tank.

The diameter of the normal vent is dependent on the tank volume and the liquid's flash point temperature. Tanks storing liquids with high vapor pressures, such as flammable liquids, require a larger-diameter normal vent than tanks storing combustible liquids. Rather than using the contents to specify the diameter of the normal vent, manufacturers assume aboveground tanks will contain Class I liquids and size the normal vent accordingly.

Table 9 Minimum Normal Vent Diameters

Tank volume (gal)	Minimum diameter (nominal pipe size, in.)
Less than 2,500	1^{1}§
2,500±3,000	1^{1}§
3,001±10,000	2
10,001±20,000	2^{1}§
20,001±35,000	3
35,000±50,000	4

Source: Ref. 16.

A properly sized normal vent will prevent overpressurizing of the tank during filling and prevents the development of negative pressure inside the tank when product is removed. Table 9 contains the normal vent sizes for shop-fabricated aboveground storage tanks [16].

The model fire codes and NFPA 30 require that normal vents for tanks storing Class I, II, and III-A liquids be located outside of buildings at least 12 ft above grade (Fig. 13). Vents cannot be located in areas where vapors may collect, such as roof eaves. If an aboveground storage tanks contains a Class I-A flammable liquid, a pressure-vacuum vent that remains closed (except when the tank is venting) is required. Tanks storing Class I-B and I-C flammable liquids require either a pressure-vacuum vent or a flame arrestor. Tanks storing Class II or III combustible liquids do not require pressure-vacuum vents or flame arrestors.

2. Emergency Vents

The most important safety feature on any aboveground storage tank is the emergency vent. All listed ASTs are manufactured with an opening to accommodate the emergency vent. Another acceptable, but less common, method is to provide emergency venting by construction. Emergency vents are provided to relieve excessive internal pressure that may be generated when a tank is heated by to an exposure or pool fire. When a tank and its contents are heated, the vapor pressure of the liquid is raised. When a liquid is heated to its boiling point, its vapor pressure equals the atmospheric pressure. If the pressure inside the tank exceeds the tank design pressure, the tank may fail. An emergency vent is designed to safely relieve the vapors created when the liquid is heated and prevent the internal pressure from exceeding the design pressure of the tank.

The importance of an emergency vent to the safety of firefighters and the general public cannot be overemphasized (Fig. 14). The emergency vent is the only available means to safely protect a steel aboveground storage tank from cata-

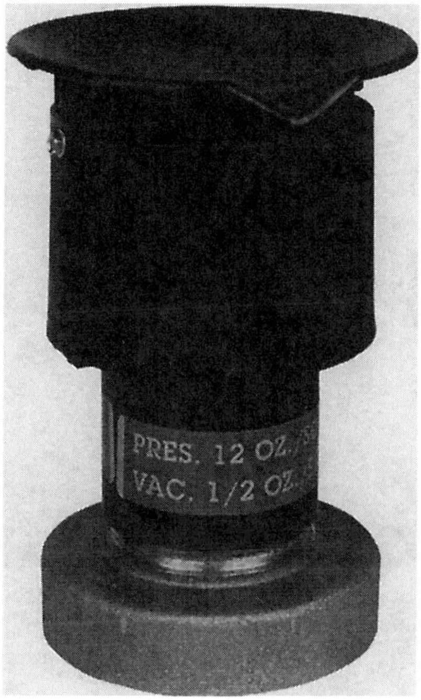

Figure 13 This pressure/vacuum vent is designed to remain closed except when product is introduced or removed from the tank. (Photograph courtesy of OPW Fueling Components.)

strophic failure. Significant pressure is created inside a steel AST when an emergency vent is disabled or improperly sized. Consider an aboveground storage tank involved in a pool fire with a disabled emergency vent. The tank contains a stable liquid that will not polymerize when heated. The pool fire heats the steel tank shell and heads. As the heat builds, the internal pressure may exceed the tank's 2.5-psig design limit. If the steel is heated to a point where it loses its strength and exhibit significant deformation (known as yield strength), or the pressure increases to a level that cannot be relieved, the tank can explosively fail. When a tank explodes, it can separate into multiple pieces. Firefighters in the immediate area of the tank may be injured or killed by the explosion overpressure or the thermal flux of the fireball.

Firefighter deaths occurred in at least two incidents because aboveground storage tank emergency vents were improperly sized or not provided. On August 18, 1959, a fire occurred at a gasoline service station in Kansas City, KS. The service station was supplied from a bulk plant on the same property. The fire involved a 21,000-gal horizontal aboveground storage tank. The storage tank did not have

Figure 14 This aboveground tank does not have an emergency vent. If the tank is subjected to a pool or exposure fire, it may explode, injuring or killing firefighters and other people nearby. This tank represents a distinct public-safety hazard and should be immediately removed from service.

a properly sized emergency vent. About 90 min into the firefighting operations, the tank catastrophically failed because of excessive internal pressure and the softening of the steel tank shell. The tank rocketed through two brick walls and landed in the street where firefighters were operating attack hose lines. Five firefighters and one spectator died in this incident. On July 31, 1968, a similar incident occurred outside of Kennedale, TX. The fire involved a home-made 10,000-gal aboveground tank storing gasoline. The tank was divided into two compartments—7000 and 3000 gal, respectively. The 7000-gal compartment was equipped with an overfill opening that acted as an emergency vent. This opening was undersized by 50 percent. The 3000-gal compartment did not have an emergency vent. Firefighters at this incident did not apply water to cool the tank because they were awaiting the arrival of firefighting foam from a neighboring fire department. Prior to the explosion, the fuel boiled at such a rate that the tank began to percolate, jumping up and down on its skids. About one hour into the incident, the tank exploded, killing two firefighters and one bystander, and injuring 57 firefighters and civilians [17]. These needless deaths and injuries would not have occurred if the tanks were equipped with properly sized and working emergency vents.

A fire official has the ultimate responsibility for determining if an emergency vent is correctly sized and functioning. Fire officials should understand how the emergency vent operates and should know how to properly calculate the required discharge flow rate. The authors of model fire codes, NFPA standards, and UL 142 have made the process of sizing an emergency vent relatively simple. Any plan review and inspection of an aboveground storage tank should include a thorough check of the emergency vent.

The model fire codes and NFPA 30 requirements for sizing emergency vents provide approximately a 4:1 safety factor, which has been incorporated into schedules that help to determine the emergency vent flow rate. These schedules were developed from fire tests performed early this century using stable liquids. However, unstable liquids were not considered when the schedules were developed in the model codes and NFPA 30. Liquids that can polymerize, decompose, condense or are self-reactive require special engineering considerations when calculating the emergency-vent size. Unstable liquids are beyond the scope of this chapter.

Correctly sizing an aboveground storage tanks emergency vent has three steps:

1. Calculation of the tank's wetted area
2. Determining the minimum emergency vent discharge flow rate
3. Ensuring that the emergency vent will adequately relieve the specified flow rate of vapor

The wetted area of a tank differs from a surface area of a tank. The wetted area actually contains liquid. When a tank is involved in fire, the liquid absorbs heat and the internal temperature rises. Eventually, vapors form and boiling occurs. The relief capacity of the emergency vent must at least equal the liquid vaporization rate, or internal pressure will continue to rise and the tank may rupture.

Pool fires are influenced by the fuel chemistry, the direction of the radiant energy, wind speed, and direction. The model fire codes and NFPA 30 require only that 75 percent of a horizontal tank's exposed surface area be considered when sizing the emergency vent. The value is based on an assumption that 75 percent of the tank contents (e.g., the tank's wetted area) will be heated by optically thick flames, which are so intense that energy radiated into the tank cannot reflect back through the flame and dissipate into the atmosphere. Research has found that such flames are only 3 to 6 ft high in hydrocarbon pool fires [18]. The 75 percent value was chosen because 100 percent of a tank's wetted area would normally not be heated in a pool fire by optically thick flames.

When calculating the emergency vent flow rate for a vertical tank, only the lowest 30 ft. of the exposed tank shell is assumed involved in the fire. Test observations found that most large hydrocarbon pool fires have flame heights ranging from 30 to 40 ft [19]. The 30-ft value was chosen on this basis.

The following examples illustrate the method for calculating the emergency vent discharge rate for horizontal and vertical tanks.

Example: Calculate the wetted area of a rectangular 2000-gal protected aboveground storage tank. The primary tank dimensions are shown in Figure 15.

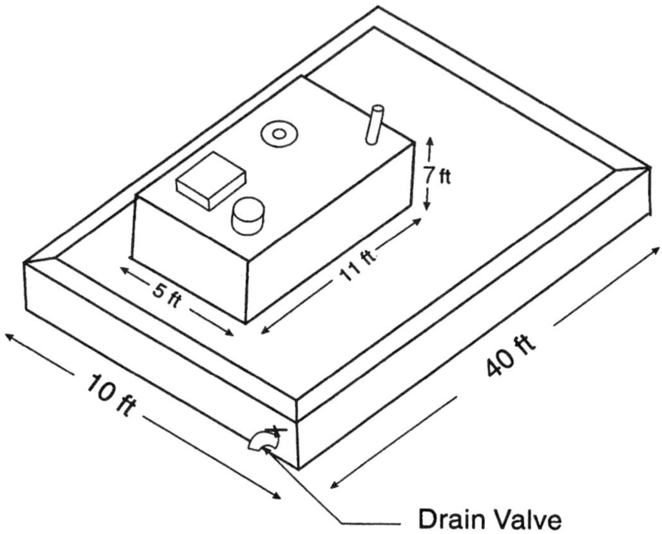

Figure 15 A rectangular 2,000-gal. protected aboveground storage tank.

$l = 5$ ft

$w = 11$ ft

$h = 7$ ft

$SA_{rectangular} = 2[(lw) + (lh) + wh)]$

$SA_{rectangular} = 2[(5\text{ ft} \times 11\text{ ft}) + (11\text{ ft} \times 7\text{ ft}) + (5\text{ ft} \times 7\text{ ft})]$

$SA_{rectangular} = 334\text{ ft}^2$

The wetted area of a horizontal tank is calculated based on 75 percent of the exposed tank surface area.

$334\text{ ft}^2 \times 0.75 = 250.5\text{ ft.}^2$

Note: Underwriters Laboratories takes a different approach to calculating the wetted area of rectangular tanks. UL does not require the area at the top of the tank to be included in the wetted area calculations. This method is more conservative than the using 75 percent of the tank surface area.

Example: Calculate the wetted surface area of a noninsulated 8000-gal vertical aboveground storage tank. The primary tank dimensions are shown in Figure 16.

$h = 29$ ft

$r = \frac{1}{2}$ the diameter of the tank, or 0.5×7 ft = 3.5 ft radius

$SA_{vertical} = 2\pi r\,(r + h)$

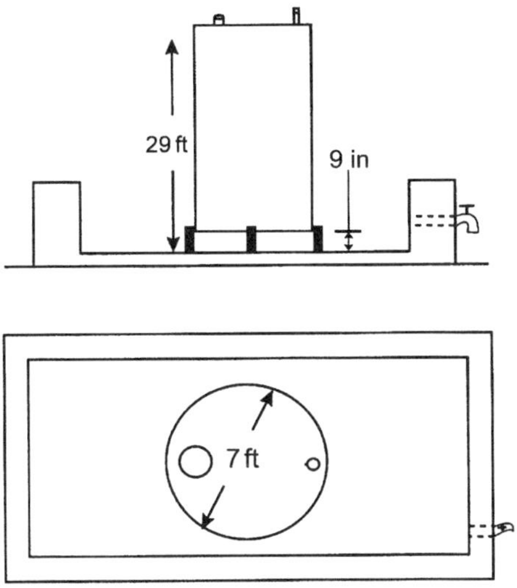

Figure 16 8000-gal vertical aboveground storage tank.

$$SA_{vertical} = 2 \times 3.14 \times 3.5 \text{ ft } (3.5 \text{ ft} + 29 \text{ ft})$$
$$SA_{vertical} = 22 \text{ ft } (32.5 \text{ ft})$$
$$SA_{vertical} = 715 \text{ ft}^2$$

After calculating the tank's wetted area, the emergency vent discharge-flow rate is determined. Table 10 contains minimum discharge-flow rates and diameters for emergency vents based on the wetted area of tanks [20]. To determine the required emergency vent flow rate, the wetted-area value is compared to the vent-capacity value shown in the second column. The specified value is the minimum required discharge capacity for the emergency vent.

For example, the wetted surface area of a 2000-gal protected aboveground storage tank is 250.5 ft^2. What is the required discharge capacity of the emergency vent?

Locate the tank's wetted area in the first column of Table 10. The adjusted wetted area of the tank is about 250 ft^2. The second column indicates the emergency vent requires a minimum discharge capacity of 239,000 ft^3/hr at 14.7 psi and 60°F. Proceeding across the row, the third column states a minimum 6-in.-diameter opening is required to safely discharge 239,000 ft^3/hr of heated vapor.

The wetted area of the 8000-gal noninsulated vertical tank is 715 ft^2. What is the required discharge capacity of the emergency vent?

Locate the tank's wetted area in the first column. The wetted area of the tank is about 715 ft^2. Interpolating the second column finds that the emergency vent requires a minimum discharge capacity of 433,100 ft^3/hr at 14.7 psi and 60°F. Pro-

Table 10 Emergency Venting Capacity for Primary Tanks and the Interstitial Space of Secondary Containment Tanks

Wetted area (ft^2)[a,b]	Venting capacity (ft^3/hr)[c,d]	Minimum opening (nominal pipe size, in.)[e]
20	21,100	2
30	31,600	2
40	42,100	3
50	52,700	3
60	63,200	3
70	73,700	4
80	84,200	4
90	94,800	4
100	105,000	4
120	126,000	5
140	147,000	5
160	168,000	5
180	190,000	5
200	211,000	6
250	239,000	6
300	265,000	6
350	288,000	8
400	312,000	8
500	354,000	8
600	392,000	8
700	428,000	8
800	462,000	8
900	493,000	8
1,000	524,000	10
1,200	557,000	10
1,400	587,000	10
1,600	614,000	10
1,800	639,000	10
2,000	662,000	10
2,400	704,000	10
2,800 and over	742,000	10

Note: Emergency venting capacity is based on atmospheric pressure of 14.7 psi and 60°F.

[a]Interpolate for intermediate values.

[b]For SI units, $m^2 = ft^2 \times 0.093$.

[c]These values are taken from NFPA 30. See NFPA 30, Section 1.2.

[d]For SI units, $m^3/s = ft^3/hr \times 0.03$.

[e]These pipe sizes apply only to open vent pipes of the speciÆc diameter not more than 12 in. long and a gage pressure in the tank of not more than 2.5 psi. If a tank is to be equipped with a venting device or Øame arrestor, the vent opening is to accommodate the venting device or Øame arrestor sized in accordance with Column 2 of this table.

Source: Ref. 2.

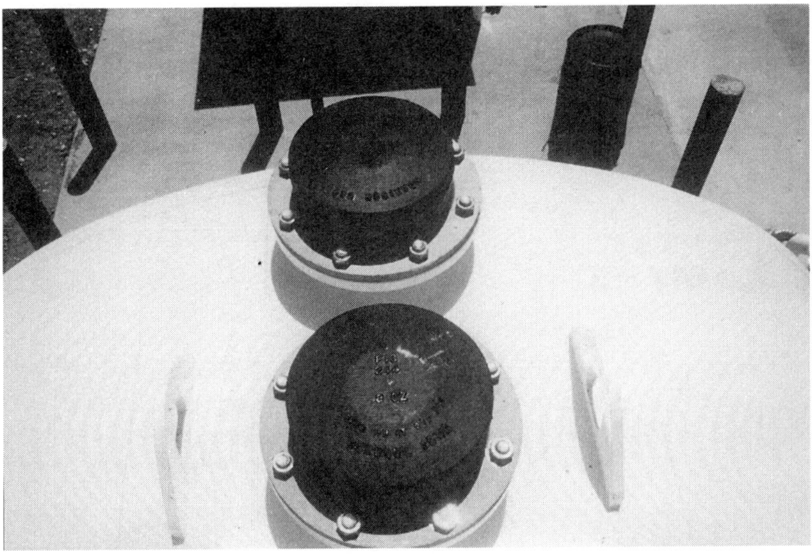

Figure 17 This 12,000-gal aboveground storage tank is constructed with integral secondary containment. The tank has two emergency vents.

ceeding across the row, the third column requires a minimum 8-in. diameter opening to safely discharge 433,100 ft^3/hr of heated vapor.

Listed aboveground storage tanks are manufactured with one or two openings for emergency vents (Fig. 17). Single-wall ASTs normally are constructed with only one emergency vent. Aboveground storage tanks with integral secondary containment commonly use two emergency vents—one of which is installed in the primary tank and the other connects to the secondary containment tank. A secondary containment tank has a larger volume than the primary tank. The larger volume is provided so that the entire contents of the primary tank can be contained. Therefore, the emergency-vent discharge rate of the secondary containment tank will be at least as large as the primary tank emergency vent.

Different types of emergency vents are installed on listed, shop-fabricated aboveground storage tanks. Many manufacturers use direct-acting vents (Fig. 18). Their principle of operation is based on the weight of vent cover, or pallet. When the pressure on the seat of the pallet equals the set pressure of the vent, the vent will begin to open. The emergency vent remains opens until pressure is relieved. After the excess pressure is relieved, the pallet closes. Direct-acting vents are typically set at the factory to operate at pressures lower than 2.5 psig. The set pressure is not adjustable in the field. The operating pressure for an emergency vent must be higher than the operating pressure of a normal vent.

Figure 18 This tank is equipped with a direct-action emergency vent. The vent is immediately left of the clock liquid-level gauge.

Long-bolt emergency vents may be used. Long-bolt emergency vents are flange cover plates installed over a manway opening. The bolts are at least 1 ½ in. in length. Bolts are typically installed in every other bolt hole. This vent is designed to lift open when the tank design pressure is exceeded.

Long-bolt vents can be easily disabled. The original bolts can be removed and the cover tightened to the manway flange. This commonly happens after a heavy rain. If the flange cover is warped or not aligned with the flange seat, water enters between the cover and seat, contaminating the tank contents. Figure 19 illustrates a disabled long-bolt emergency vent. Fire officials should closely scrutinize long-bolt vents and should make every effort to convince the tank owner to use a more reliable type of emergency vent.

NFPA 30 and the model fire codes require permanent markings on emergency vents that indicate their discharge capacity, which is expressed in standard ft^3/hr (SCFH) at 2.5 psig. During an inspection, the discharge capacity of the emergency vent should be compared to the tank nameplate data. The discharge rate of the vent must at least equal the discharge rate indicated on the tank nameplate. Some tank manufacturers use the combined flow rates of the normal and emergency vents to accommodate the required emergency-vent discharge capacity. This is permissible if the discharge capacity of the normal vent is marked on the device.

API Standard 2000 allows other types of emergency vents for aboveground storage tanks including:

Figure 19 This aboveground storage tank is equipped with a long-bolt emergency vent. The bolts on the vent have been tightened, disabling the emergency vent. This tank presents a distinct hazard to the safety of firefighters and the general public. Repairs to the vent should be made immediately, or the tank should be removed from service.

A gauge hatch that permits the cover to lift when abnormal internal pressure develops.

A connection between the roof and shell that is weaker than the weakest vertical joint in the shell or the shell-to-bottom connection. A tank with a frangible roof-to shell attachment does not require emergency vents. Care should be taken to ensure that the API 2000 requirements for a frangible roof-to-shell attachment are met, particularly on a smaller tank, before this method of emergency venting is used.

Other forms of construction that can be proven comparable for the purposes of pressure relief.

A rupture disk device [21].

H. Overfill Prevention Systems

Preventing tanks from being overfilled is an important safety and environmental consideration. The safe operation of aboveground storage tanks is dependent upon the receipt of product into the intended storage tank within its defined capacity. Tank overfills can be effectively reduced by developing and implementing practical, safe operating procedures for storage facilities and by providing for careful selection and application of equipment, scheduled maintenance programs, and employee training [22].

In reviewing AST fires, NFPA found that a common factor in many incidents was tanks being overfilled. Once a liquid is outside of a tank or piping system, it begins to form a flammable vapor cloud. If a source of ignition is present, the vapor cloud can ignite, causing a liquid-pool fire or vapor-cloud explosion. United States Coast Guard data in Table 6 indicates the number of spills occurring for unknown reasons is increasing. Many of these incidents were probably the result of tank overfills. The cause for the overfilling is most likely human error.

Data on aboveground storage tank incidents from the American Petroleum Institute also reinforces the need for overfill protection. The 1988 API report prepared by Events Analysis, Inc. reviewed 130 AST spill incidents. Table 11 summarizes the findings.

Model fire codes and NFPA standards have requirements for aboveground storage tank overfill protection systems. These requirements vary based on the intended aboveground storage tanks' use and its contents. The model code requirements are intended to minimize or prevent overflows that can result in environmental and safety hazards, loss of inventory, and damage to tanks and adjacent areas (Table 12).

An overfill prevention system should stop the delivery of liquid into the tank (Fig. 20). By itself, a high-liquid-level alarm does not constitute an overfill-prevention system because it is not designed to stop liquid flow into the tank. An alarm may inform a person that the tank is full; however, it cannot stop the person from adding more product and overfilling the tank. If a high-liquid-level alarm is connected to a pump or solenoid valve and engineered so a pump stops or a valve

Table 11 Causes of Aboveground Storage Tank Leaks

Cause	Percentage of reviewed incidents
Vandalism	19.6%
OverÆlls	15.2%
Valve failures	9.8%
Maintenance	6.5%
Rupture	5.4%
Corrosion	4.3%
Damage due to vehicle impact	4.3%
Floods	3.3%
Lighting	3.3%
Pipe leaks	2.2%

Source: American Petroleum Institute.

Table 12 Typical Model Fire Code Requirements for Tank OverÆll Prevention

Application	Typical model Ære code requirements
Motor vehicle fueling	Tanks used for motor vehicle fueling require an alarm that activates when the tank is Ælled to a predetermined level (85% or 90% of its rated volume) and an automatic method of stopping the Øow at 90% or 95%.
Petroleum bulk plants or terminals	Aboveground storage tanks at terminals receiving Class I liquids from mainland pipelines or marine vessels require a method of preventing overÆlls. The model codes reference the requirements in API 2350 (Overfill Protection for Storage Tanks in Petroleum Facilities).
Tanks inside of buildings	A means of preventing tanks inside of buildings from overØowing liquid is a requirement of all the model codes. NFPA 37, Chapter 6 requires overflow protection for indoor day tanks supplying fuel to stationary combustion engines.
Other applications using aboveground storage tanks storing Class I, II, and III-A liquids	NFPA 30 allows the use of tanks with integral secondary containment in lieu of constructing a dike area. Conditions of approval require an alarm that sounds at 90% of the tank capacity. The Øow must stop when the liquid level reaches 95% of the tank capacity.
	The BOCA, Southern, and Uniform Fire Codes require a high-level alarm on hazardous materials storage tanks larger than 500 gal capacity. Some jurisdictions apply this provision to ASTs containing Class I, II, and III-A liquids.

closes at a preset level, the arrangement should be acceptable as an overfill prevention system.

I. Liquid Piping

Piping is commonly used for moving liquid into and out of a tank to an application. The model fire codes and NFPA 30 contain requirements for the design, construction and testing of underground and aboveground piping. They also reference American Society of Mechanical Engineers requirements in ASME B31.3 (Chemical Plant and Petroleum Refinery Piping). This code prescribes requirements for the materials, design, fabrication, assembly, erection, examination, inspection, and testing of all piping within the property limits of facilities engaged in the processing or handling of chemicals, petroleum or related products. The ASME B31.3 requirements apply to piping that handles raw, intermediate, and finished chemical and petroleum products.

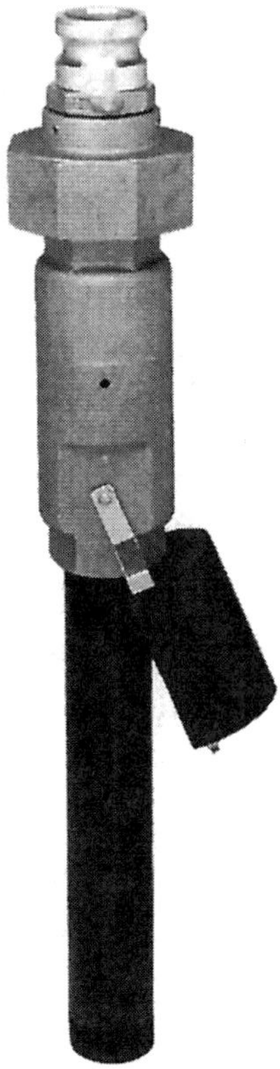

Figure 20 An overfill prevention device is required for aboveground storage tanks used for motor vehicle fueling. Model codes and standards specify overfill protection for other applications. (Photograph courtesy of OPW Fueling Components.)

The requirements in the model codes and standards offer sufficient design guidance for aboveground piping used for most motor-vehicle fueling applications. The selected schedule and specification of pipe used should meet the material requirements of ASME B31.3. For example, low-melt-point materials such as aluminum, copper and nonductile materials are allowed for Class I, II, and III-A liquids. However, they must be adequately safeguarded from fire exposure, buildings and persons. Adequate safeguarding may require a fire-resistant coating applied to pipe and fittings, additional remote valves so piping can be isolated if a fire occurs, or locating the pipe in area where a leak or fire will not impact persons and buildings. For motor vehicle fueling, NFPA 30A permits the use of these and other low-melt-point materials when pipe and fittings are buried.

Valves used in aboveground piping systems should be constructed of steel or nodular iron meeting ASTM A 395 (Ferritic Ductile Iron Pressure-Retaining

Castings for Use at Elevated Temperatures). A sufficient number of valves are required to operate the system properly and to protect the facility surrounding the tank. For motor vehicle fueling, this normally requires valves at the tank opening, at the discharge side of the transfer pump and, at the base of the dispenser. A method of hydrostatic relief to protect the piping from the pressure that can be developed by liquid trapped between two closed valves is required.

Antisiphon valves are required when piping is connected at the top of the tank and extends below the liquid line. It is designed to prevent the siphoning of liquid if the pipe leaks or fails. Electric or pneumatic solenoid valves that open on demand are commonly used for this purpose. Figure 21 illustrates how an antisiphon valve is installed on an AST. Figure 22 shows a cross section of an antisiphon valve.

Pipe and fittings require adequate support and protection against physical damage and stresses induced by soil settlement, expansion, contraction and vibration (Fig. 23). If piping supports are located in an area where risk of fire exposure is high, protection in the form of drainage, fire-resistant coatings, or water spray systems is necessary.

Piping must be protected against corrosion, which can deteriorate the pipe wall material due to an electrochemical reaction between the pipe and its environment. To protect against corrosion, engineering design normally provides for additional material in the pipe wall, the use of a suitable coating or lining, or the specification of a resistant material. The method used to deal with corrosion depends upon the corrosion rate of the soil and atmosphere around the piping.

For steel pipe, corrosion protection can be provided using cathodic protection and external coatings—either applied at the factory or in the field. The design

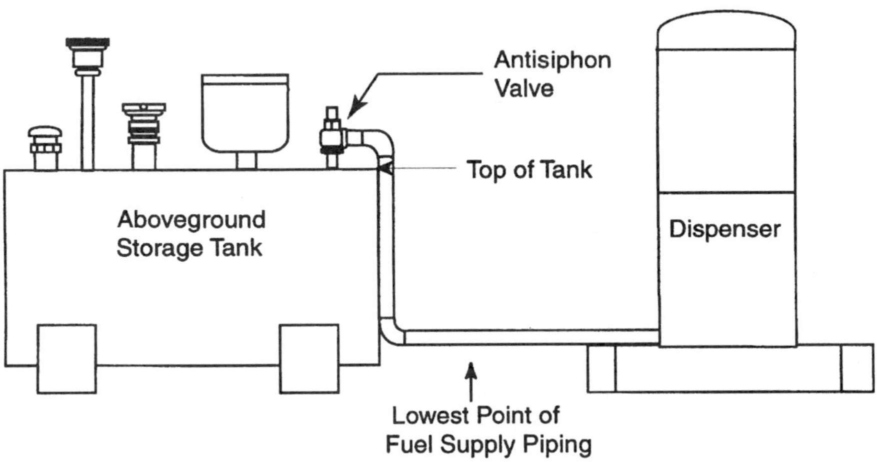

Figure 21 Illustration of AST with antisiphon valve.

Figure 22 Antisiphon valves are designed to prevent liquid siphoning if piping or fittings below the tank liquid level leak or fail. (Photograph courtesy of OPW Fueling Components.)

and installation of corrosion protection for steel pipe should be in accordance with National Association of Corrosion Engineer (NACE) International Recommended Practice 01-69 (Control of External Corrosion on Underground or Submerged Metallic Piping Systems).

Plastic pipe and fittings can be attacked by solvation, a chemical reaction that penetrates the plastic via a corrosive process that causes softening, swelling, and ultimate failure. Plastics, in contrast to metals, do not exhibit a corrosion rate; usually they either completely resist attack or deteriorate rapidly.

When corrosion is anticipated to occur at a slow, regulated rate and this rate can be reliably predicted, the usage of a corrosion allowance—providing excess material in the pipe wall—often is a preferred solution. This material will be consumed over the design life of the piping system and therefore cannot be counted upon to serve any other purpose such as pressure integrity or mechanical strength [23]. NFPA 30 requires basing the protection to be employed upon the corrosion history of the area and the judgment of a qualified engineer. The authority having

Figure 23 Piping must be adequately supported and protected against physical damage. This pipe is supported using a concrete footing and an anchoring system made of corrosion-resistant materials.

jurisdiction can waive the requirements for corrosion protection after qualified engineer demonstrates that such protection is not necessary [24].

A registered professional engineer should supervise the design of a piping system for flammable or combustible liquids. This is necessary because ASME B31.3 is an engineering code that requires an understanding of materials, fabrication, assembly, inspection, and testing methods.

ASME B31.3 does not specify the type of pipe or tubing that is required for an application. This is the responsibility of the designer. When a piping system is designed, key considerations for evaluation are:

Possible exposure to fire with respect to the loss of strength, degradation temperature, melting point, or the combustibility of the pipe or support material

Ability of thermal insulation (when required) to protect the pipe from fire

Susceptibility of the pipe to brittle fracture, possibly resulting in failure from thermal shock when exposed to fire or firefighting measures

Susceptibility of the piping material to crevice corrosion in screwed joints, or adverse electrolytic effects if the metal is subject to contact with dissimilar metals

Suitability of packing, seals, gaskets, and lubricants or sealants used on threads as well as compatibility with the fluid handled [25]

ASME B31.3 divides piping into several different service categories. Three fluid services that are of concern to fire officials are Category D, normal, and Category M. Category D fluid service is defined as "a fluid service to which all of the following apply":

The fluid handled is nonflammable and nontoxic
The design gauge pressure does not exceed 150 psi
The design temperature is between -20°F and 360°F

Category M is defined as "fluid service in which a single exposure to a very small quantity of toxic liquid, caused by leakage, can produce serious, irreversible harm to persons on breathing or body contact, even when prompt restorative measures are taken."

For most flammable and combustible liquids, the requirements for normal fluid service apply. Normal fluid service is "pertaining to most piping covered by this code, i.e., not subject to the rules for Category D, Category M, or high-pressure fluids service, and not subject to severe cyclic conditions [26]."

Although ASME B31.3 is an engineering code, it does contain important requirements and limitations that plans examiners and inspectors should understand. These were developed from the committee's understanding of materials, fabrication and installation methods. Some of the ASME B31.3 requirements and limitations for flammable and combustible liquids in normal fluid service include:

Furnace lap-weld ferrous pipe, furnace butt-weld ferrous pipe, spiral-weld ferrous pipe, and fusion-welded steel pipe (made to ASTM A134, A53 Type F, API 5L furnace butt welded, and A211) are not permitted for hydrocarbons or other flammable liquids. Nonferrous pipe made by similar manufacturing processes is also restricted [27].

Pipe used for normal fluid service that is assembled with threaded joints must have a minimum wall thickness equal to Schedule 40. Depending on the pipe material, Schedule 80 piping may be required. Threaded joints on pipe over 2 in. in diameter used for flammable fluids service is not allowed unless the joints are safeguarded [28]. Some methods of safeguarding include drainage to a safe location to prevent liquid from accumulating under pipe supports, water spray systems, or use of fire-resistant materials to insulate the fittings. Additional guidance can be found in Chapter 3 of NFPA 30.

Pipe and materials constructed of ductile iron cannot be repaired or assembled by welding. Pressure-containing parts made of cast iron are not allowed for flammable liquid service operating at temperatures greater than 300°F or 150 psig. Pressure-containing parts that contain high silicon iron (14.5 percent) are prohibited from use in flammable fluid service [29].

Steel pipe welders should be qualified in accordance with American Welding Society standards. ASME B31.3 also specifies requirements for pipe assembled by brazing, fusion welding, chemical fusion and other methods.

When a compound or lubricant is used on threaded connections, the material must be suitable for the service conditions and not react unfavorably with either the fluid or the piping material. For flammable and combustible liquid service, pipe compounds should be listed to UL Subject 1356, (Outline of Investigation of Pipe Joint Sealing Compounds). The document evaluates pipe joint-sealing compounds based on the type of fluid, the operating pressure of the pipe and its diameter.

ASME B31.3 prohibits the use of thermoplastic pipe for aboveground flammable liquid service. Piping constructed of reinforced plastic mortars, plastic thermosetting resins, borosilicate glass and porcelain may be used for flammable fluids service when properly safeguarded. This provision does not eliminate using these materials in aboveground applications when they are used as an internal lining for pipe constructed of materials allowed for normal fluid service.

J. Electrical Equipment

NFPA 70, the National Electrical Code® (NEC®),* is adopted by local, provincial, and state electrical enforcement agencies as the minimum standard for the safe design and installation of electrical equipment. The NEC® contains requirements for electrical equipment used in areas designated as classified or hazardous locations. The intent of the NEC® is to prevent flammable vapors from being ignited by electrical equipment and wiring.This designation can be assigned to an area or space when a fire or explosion hazard may exist due to the presence of flammable gases, flammable liquids or their vapors. Classification of an area is not necessary unless the flammable gas or vapor may be present in the air in explosive or ignitable quantities. Fire officials can require the installation of hazardous (classified) location electrical equipment in and around tanks, pumps and dispensers for flammable liquids.

Several variable conditions influence when an aboveground storage tank and the area around it require hazardous (classified) electrical service. The classification of a hazardous location requires an evaluation of a material's properties, the environment in and around the point where liquids or vapors may be released, and the probability of flammable vapors escaping from the process. Conditions that influence the classification include:

A material's flash point and autoignition temperature

---

*National Electrical Code and NEC are registered trademarks of the National Fire Protection Association, Inc., Quincy, MA 02269.

Operations in or near the flammable range
Process temperatures above the liquid's flash point or boiling point temperatures
Process temperatures above the material's autoignition temperature

The NEC® requirements for flammable liquids consider the flashpoint and autoignition temperature, the operating temperature of the equipment and whether flammable vapors are present under normal operating conditions or in case of equipment breakdown or mechanical repairs. Hazardous (or classified) location electrical equipment is not required in and around areas storing or using class II or III combustible liquids unless they are heated above their flash-point temperature.

Classification of hazardous locations is based on the type of material in a location, the hazard category of the material, and the potential for the material to be in a concentration that is ignitable. The NEC® specifies the requirements for hazardous (classified) locations based on the:

Class of material
Hazard group of the material
Degree of hazard, or division

Therefore, a properly classified hazardous location will have a class, group, and division.

The NEC® designates flammable liquids as Class I materials. It should also be noted that the liquid "classes" are not related to the hazardous location "classes." The NEC® further subdivides materials using the classification system in NFPA 497M (recommended practice for the classification of flammable liquids, gases or vapors and of hazardous [classified] locations for electrical installations in chemical process areas). The material classification is assigned according to a liquid's flash-point and auto-ignition temperatures. For Class I materials, NFPA 497M subdivides them into Groups A, B, C, and D. This subdivision is based on the material's propensity to ignite inside of a test apparatus. Generally, the ease of ignition is relative to the group designation. In other words, a Group A material is more easily ignited inside the test apparatus than a Group D material. Many Class I-B and Class I-C flammable liquids are designated as Group D materials by NFPA 497M.

The NEC® also has requirements for the type of electrical equipment based on the potential presence of flammable vapors in the atmosphere. Class I materials can require equipment designed for Division 1 or Division 2 locations. In a Division 1 location, an ignitable mixture is constantly or intermittently present under normal operating conditions, repair, leakage or during maintenance. An example of a Division 1 location is the atmosphere inside an AST filled with gasoline. Vapors are constantly present under normal operating conditions. In a Division 2 location, an ignitable mixture is likely only under abnormal conditions, such as the

Table 13 NEC® DeÆnitions for Class I, Division 1 and 2 Locations

Class I locations	Requirements
Division I location	A location (1) in which ignitable concentrations of Øammable gases or vapors can exist under normal operating conditions; (2) in which ignitable concentrations of such gases or vapors may exist frequently because of repair or maintenance operations or because of leakage; or (3) in which breakdown or faulty operation of equipment or processes might release ignitable concentrations of Øammable gases or vapors and might cause simultaneous failure of electrical equipment that could act as a source of ignition.
Division 2 location	A location (1) in which volatile Øammable liquids or Øammable gases are handled, processed, or used, but in which the liquids, vapors, or gases will normally be conÆned within closed containers or closed systems from which they can escape only in case of accidental rupture or breakdown of such containers or systems, or in case of abnormal operation of equipment; or (2) in which ignitable concentrations of gases or vapors are normally prevented by positive mechanical ventilation and might become hazardous through failure or abnormal operation of the ventilating equipment; or (3) that is adjacent to a Class I, Division I location and to which ignitable concentrations of gases or vapors might occasionally be communicated unless such communication is prevented by adequate positive-pressure ventilation from a source of clean air and effective safeguards against ventilation failure are provided.

Reprinted with permission from NFPA 70-1996, the *National Electrical Code*®, Copyright © 1995, National Fire Protection Association, Quincy, MA 02269. This reprinted material is not the complete and ofÆcial position of the National Fire Protection Association, on the referenced subject which is represented only by the standard in its entirety.

failure of a seal on a flammable liquid pump. Table 13 summarizes the NEC® Division 1 and 2 location requirements [30].

The NEC® has requirements for hazardous-location electrical equipment installed at aboveground storage tanks and gasoline service stations. NEC® Article 514 specifies hazardous-location requirements for gasoline dispensing and service stations, and Article 515 specifies requirements for bulk storage plants. The requirements designate the boundaries for Division 1, Division 2, and nonclassified locations around dispensers and aboveground storage tanks. The requirements also address acceptable wiring (conductors) types and installation practices, arrangement of circuit disconnects and conduit seals, grounding, and the location of dispenser emergency shutoff switches.

K. Liquid Pumps

Liquids are commonly transferred into and out of tanks by gravity or pumps. Systems utilizing gravity flow eliminate the expense of purchasing and operating a pump. Gravity flow of liquids can be hazardous if the transfer system is not properly designed. If valves are not appropriately installed so the flow can be safely and effectively controlled, the entire tank contents may be released. The accidental release of the tank contents can occur when pipe, fittings or valves are located below the tank's liquid line.

A preferred method of transferring liquids is pumping from the tank to its end use. Pumps can be arranged to manually or automatically start and stop on demand. Pumps may also be installed so if a high liquid-level switch activates, the pump is stopped. This reduces the potential for an overfill. When installed on top of a tank, the pump must take suction or be submerged in the liquid. This design virtually eliminates the potential for an uncontrolled loss of the tank contents. When installed below the tank liquid line, pumps serve as an in-line method of controlling liquid flow.

UL 142 steel tanks can be arranged with fitting openings above or below the liquid level. Fitting openings are located on top of listed protected and fire-resistant tanks (Fig. 24). UL 2080 and UL 2085 prohibit openings below the liquid line.

Transfer pumps for flammable and combustible liquids should meet UL 79 (Standard for Power-Operated Pumps for Petroleum Dispensing Products). UL 79 contains requirements for electric, hydraulic and pneumatic-powered pumps for

Figure 24 This tank is equipped with two listed submersible pumps used for delivering fuel oil to generator sets. These pumps have a maximum discharge pressure of 50 psig, as required by UL 79.

use with gasoline/alcohol blends, diesel fuel, fuel oil and lubricating oil. Other petroleum products are not prohibited if the pump was evaluated for the liquids in question and it is marked with the specific liquid name.

Self-contained dispensers and submerged pumps used with remote-control liquid dispensers are limited to discharge pressures of 50 psig. The discharge pressure for pumps that transfer liquids from one tank to another vehicle or tank are not limited by UL 79. Pumps with discharge pressures greater than 50 psig are equipped with a means of pressure relief. This may be accomplished using factory-sealed relief valves located inside the pump housing or bypass lines.

UL 79 requires pumps be subjected to various performance and safety tests. These tests evaluate the pump for leakage, hydrostatic durability, compatibility of materials that may be in contact with liquid or vapors, and the pump's endurance under normal and maximum discharge pressure conditions. Other tests are performed depending on the type of power supply used to drive the pump.

Requirements for pumps in the model codes and NFPA standards can vary based on the application. Positive displacement pumps handling liquids require a means of pressure relief that discharges to a suitable location. A suitable location may be into a tank or a bypass line that returns the discharge to the pump-suction connection. Listed pumps are required when used with motor vehicle fueling dispensers.

L. Dispensers

The dispenser is the point at which the vehicle operator has direct access to the fuel. Motor vehicle fuels are hazardous materials. Therefore, fire officials should ensure that the appropriate type of dispensers and equipment are specified in the design.

NFPA 30A and Article 52 of the 1997 UFC require the use of listed dispensers that meet UL 87 (Standard for Power-Operated Dispensing Devices for Petroleum Products). UL 87 defines a dispenser as "a product consisting of a meter, motors, or fluid control, and an area for storing a hose-nozzle valve with or without a pump." A dispenser can be self-contained or remotely controlled. The primary difference between the dispenser types is the pump location. Self-contained dispensers commonly have a suction pump at the dispenser [31]. Remote control dispensers do not contain a pump.

Normally, the electrical code official and fire official designate classified locations. UL 87 contains extensive requirements that specify the boundaries of Division 1, Division 2, and nonclassified atmospheres inside the dispenser. Because designs vary between manufacturers and technology continues to change, NEC® Table 514-2 states that boundaries for hazardous locations inside the dispenser will be established by nationally recognized testing laboratories. For example, the area where the dispenser is connected to the piping is normally designated a Class I,

Group D, Division 1 location. However, the area inside the dispenser may be Division 2 if the electronic fuel delivery display is separated from the pump by a vapor-tight seal. This requirement does not supersede the code official's authority to establish classified location areas. It is intended to alleviate the burden of evaluating each type or style of dispenser. The provision gives manufacturers the flexibility to apply designs that can de-rate an enclosure to a Division 2 or nonclassified location. Figure 25 illustrates the Division 1, 2, and nonclassified areas of dispensers based on the UL 87 requirements.

Listed valves are required with motor vehicle fuel dispensing and such fuel-burning equipment as diesel-powered generator sets. For these applications, several NFPA standards and the model fire codes require the installation of listed valves that meet UL Standard 842 (Valves for Flammable Liquids). The standard contains testing requirements for emergency shutoff, fusible link and hose-nozzle valves [32].

NFPA 30A and the model fire codes require the installation of listed emergency shutoff valves in remote and overhead dispensers (Fig. 26). An emergency shutoff valve is designed to stop the flow of fuel in case of a fire at or near the dispenser. The valve is also designed to automatically stop fuel flow if the dispenser is sheared off or pulled away from its foundation. Emergency shutoff valves are required at the base of remote dispensers and at the inlet of overhead dispensers.

Most emergency shutoff valves are designed to operate as spring-loaded shutoff valves. The valve is maintained open by a low-melt-point fusible link, typically rated at 165°F or 212°F. If the valve is subjected to a pool or exposure fire,

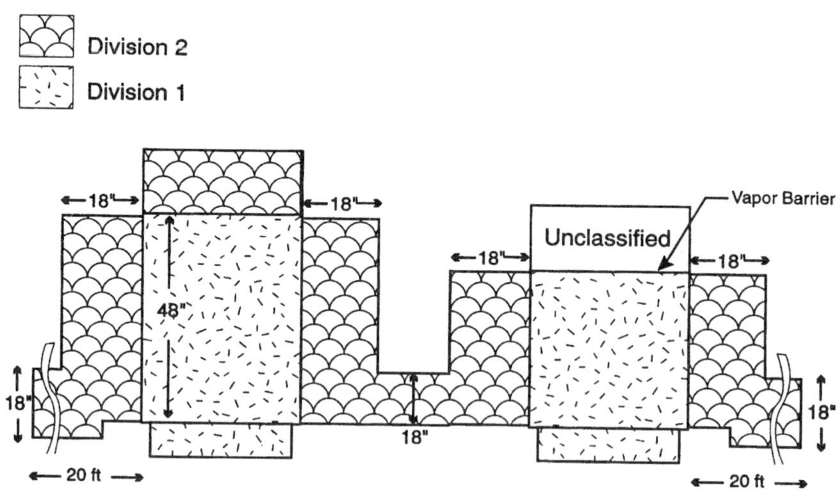

Figure 25 The division 1 and division 2 location boundries for dispensers are specified in NEC® article 514 and UL 87.

Figure 26 An emergency shutoff valve. (Photograph courtesy of OPW Fueling Components.)

the link melts, which closes the valve. The valve also incorporates a horizontal shear section. The shear section is installed parallel with the fuel island or foundation. If the dispenser is knocked over or pulled away from its foundation, the shear section separates from the valve body, causing the valve to close.

Before emergency shutoff valves are listed, they must successfully complete a series of tests specified in UL 842. Listed valves are also subjected to tests during the manufacturing process.

UL 842 does not allow liquid leakage past the valve seat when it is subjected to any pressure between zero and 1.5 times its maximum rated pressure. The valve is also subjected to a fire test to determine if it will close when exposed to a pool fire, and limit the leakage of flammable liquid that would prolong the blaze. During the fire test, the valve must meet its rated operating pressure. After 30 min of fire exposure, the pool fire is allowed to burn out. After 45 min, the valve is inspected for leaks that may contribute to an external fire.

Another test evaluates the valve shear section. The valve must close when subjected to a bending moment of no more than 650 lb-ft at the connection to the pipe or flange. Following this test, the valve is examined for external and seat leakage.

Listed nozzle valves are required for fueling motor vehicles. Automatic hose nozzles are required to incorporate a means that shuts off liquid flow during fueling to prevent tank overflow from the vehicle fill opening [33]. Hose nozzle valves are evaluated using UL 842.

The NFPA 30A and model fire code requirements also consider the person dispensing the fuel. Dispensing is classified as either attended or unattended. Attended dispensing requires the presence of a person whose primary responsibility is ensuring the safe transfer of liquid from the dispenser into the vehicle. Unattended dispensing allows the general public to control the dispensing operation. Many service stations today are convenience stores that also sell flammable and combustible liquids. The person in attendance is concerned about selling a variety of goods and services besides gasoline or diesel. Accordingly, most fire officials treat dispensing of fuel at convenience stores as unattended. This can require additional controls beyond those mandated for unattended dispensing.

Hose nozzles may, or may not, have a latch-open device. NFPA 30A and the model fire codes require the installation of latch-open devices at unattended service stations. The provision's intent is to ensure that the person dispensing fuel does not insert a foreign object between the operating lever and the valve body, causing the valve to be blocked open. For example, a person inserts his wallet between the lever and valve body so that he can go into the store and purchase a six-pack of beer. Because the valve is forced open, liquid overflows outside the tank, causing a fuel spill. If this same person introduces an ignition source, a pool fire will occur.

UL 842 addresses hose nozzle construction and performance. Hose nozzles must close automatically upon manual or automatic release of the operating lever. When assembled, the valve requires electrical continuity from the spout to the inlet of the valve body without the use of a jumper wire. This safety feature ensures that a continuous path to the electrical ground is provided if static electricity occurs while dispensing fuel. Valves are also subjected to drop tests that evaluate their ability to withstand normal use. Automatic nozzles are subjected to drop tests to ensure the valve will automatically close if it is accidentally dropped while dispensing fuel.

Listed hoses are required on dispensers used for motor vehicle fueling. Hoses are listed using the requirements in UL 330 (Hose and Hose Assemblies for Dispensing Flammable Liquids). Listed hose and hose assemblies are designed for use on gasoline and diesel fuel dispensers. Hoses for other fuels may be used when they have been investigated for resistance to the particular fuel. These hoses are intended for use at a maximum working pressure of 50 psig.

Hose and hose assemblies constructed to the UL 330 requirements are subjected to a number of tests. These tests include hydrostatic burst strength, repeated bending of the hose, electrical continuity between the hose and couplings, and resistance to gasoline and diesel.

UL 330 was recently revised to prohibit reattachable fittings on couplings used on listed hose assemblies. A reattachable coupling can be removed from a damaged or discarded hose and reinstalled on a repaired or completely different hose, but the coupling may not be properly installed or suitable for another hose. Leakage or loss of electrical continuity between the couplings could result. Also, the listing mark for hose assemblies is often on the coupling and, unless removed, could appear on an altered or completely different hose assembly [34].

M. Protecting the Tank from Damage

The model fire codes contain provisions intended to protect aboveground storage tanks from exposures besides fires and spills. Two specific areas of interest are vehicle impact protection and projectile resistance.

The model fire codes have requirements for protecting aboveground storage tanks from vehicle impact (Fig. 27). A common method of protection is installing

Figure 27 Because the tank was located in an area where cars and trucks operated, vehicle impact protection was provided.

pipe bollards around the tank. Fire codes specify requirements for minimum diameter, spacing, height and footing dimensions. Installations designed to the code requirements should protect a tank from passenger vehicles operating at low speeds. Vehicle impact protection may be unnecessary if the tank is located where impact by vehicles is unlikely or impossible. Fire officials can modify or waive vehicle impact protection requirements.

Listed vehicle impact resistant protected and fire-resistant tanks do not require pipe bollards or other external protective means. Manufacturers can test the aboveground storage tank to determine if it can withstand a 12,000-lb load applied to the tank at a speed of 10 miles per hour. The load is applied at a point 18 in. above the tank foundation. The test load is directed to the most vulnerable side of the tank. Depending on the manufacturer's installation instructions, the tank may be restrained or unrestrained. After the dynamic load test, the tank is hydrostatically tested. If the tank successfully passes the 5.0 psig hydrostatic test, it can be listed as vehicle impact resistant.

A listed vehicle impact–resistant tank cannot withstand all forms of damage. When a vehicle strikes a tank, attached piping may be displaced. An impact that dents or creases the tank shell may induce stresses that cannot be identified visually. Since the tank is fire resistant or protected, the impact may have damaged the insulating material. For these reasons, a tank and connected piping that have been struck by a vehicle should be tested for liquid integrity. Preferably, the contents should be removed and the tank should be tested by the organization that issued its listing.

Recertification of a tank that has been struck by a vehicle will require a representative of the testing laboratory to travel to the site and examine the tank. In certain cases the tank may need to be shipped back to the manufacturer for examination. Neither of these options is inexpensive. Therefore, it is prudent to ensure that the method of vehicle impact protection will shield the tank from any reasonably foreseeable impacts. More important is that vehicle impact provisions protect persons from injury or death if they were in a vehicle that punctured a tank filled with a flammable or combustible liquid.

Fire officials may require that an aboveground storage tank be projectile resistant. This may be necessary if the tank is located in an area where stray or intentional bullets from firearms could strike the tank.

Protected and fire-resistant tanks can be listed as projectile-resistant. The test subjects the tank to gunshots from a 30-caliber rifle located 100 ft from the tank. Five rounds of 150-grain military ball ammunition are fired at the tank's most vulnerable area. A tank can be listed as projectile resistant if it passes a hydrostatic test after the ballistics test.

Noninsulated steel tanks are not listed as projectile resistant. However, protection from firearm projectiles can be provided for any aboveground storage tank. The Institute of Makers of Explosives has performed bullet-resistance tests on various building designs. The test results have been codified in federal regulations and the model fire codes. The designs were evaluated using the same test performed by the testing laboratories for evaluating the projectile resistance of ASTs. Some examples of bullet-resistant construction include:

8-in.-thick solid concrete
⅝-in.-thick steel
8-in. concrete masonry units with the interior voids filled with a well-tamped sand/cement mixture

The Southern Fire Prevention Code and NFPA 30A require a security fence around aboveground storage tanks. A minimum 6-ft-high fence is required around the tank. Fence openings must be secured against unauthorized entry. A minimum 20-ft separation is required between the fence and tank.

III. INSPECTION OF ABOVEGROUND TANKS

A. Acceptance Inspection

After an aboveground storage tank is installed, a fire official should inspect it. This inspection is necessary to verify that the installation complies with adopted codes and standards, to ensure the installation is liquid-tight, and to verify that required safety controls function as intended. Areas of focus include an examination of the tank's nameplate data, pipe pressure tests, functional tests of components, an inspection of the electrical installation, and required special inspections.

1. Tank Inspection

Nameplates are installed on all listed, shop-fabricated storage tanks (Fig. 28). They serve as a permanent record of the standard to which the tank was constructed, specific design and installation features, and who manufactured the tank. Reviewing the nameplate should be the first inspection step. The nameplate will indicate if the correct type of tank was installed. The presence of specific tank features such as integral secondary containment, vehicle impact protection and projectile resistance is indicated on nameplates.

Nameplates contain the required discharge rate for the emergency vents. The nameplate data should be compared to the discharge capacity of the emergency vent. If the value on the emergency vent is less than that specified on the nameplate, the emergency vent is undersized. A tank should not be filled with product until the size of the vent is corrected.

A tank inspection should include an evaluation of the supports to ensure they are provided in accordance with the manufacturer's or engineer's design. The inspection should visually survey for damage to the tank shell and heads. For sec-

(File number)

UL LISTED

INSULATED SECONDARY CONTAINMENT ABOVEGROUND TANK FOR FLAMMABLE LIQUIDS PROTECTED TYPE NO. ____________

Manufactured by: ____________ Model: ____________

Date of Manufacture ____________ Manufacturer's Code ____________

Primary Tank Capacity ____________ Gallons

Maximum Leakage Test: 3 PSIG

Note: This tank requires emergency relief venting capacity not less than ____C.F.H. (Primary Tank) and ______ C.F.H. (Annular Space).

Pressurize Primary Tank when Pressure Testing Annular Space.
THE TANK HAS NOT BEEN INVESTIGATED TO DETERMINE ACCEPTABILITY FOR USE AFTER EXPOSURE TO FIRE.

This tank is for stationary installation only.
VEHICLE IMPACT AND PROJECTILE RESISTANT.
This aboveground tank manufactured in accordance with: UL 142, UL 2085, and UFC 79-7.
This aboveground tank is intended for installation in accordance with UFC Appendix II-F, NFPA30, and NFPA30A.
Install approved overspill containment and overfill prevention before filling - SEE OWNERS MANUAL.

TANKS TO BE INSTALLED PER INSTALLATION INSTRUCTIONS.

Figure 28 Nameplates are provided to assist inspectors. Nameplates indicate the standard the tank was constructed to, the discharge flowrate of the emergency vents, as well as the manufacturer and date of fabrication. (Courtesy of Underwriters Laboratories Inc., Santa Clara, CA.)

ondary containment tanks, the primary tank should be pressure tested in accordance with NFPA 30 requirements.

There are manufacturers of shop-fabricated ASTs that are not listed by nationally recognized testing laboratories. Tanks from such companies may have a nameplate, but they will not have a listing mark. Fire officials should not approve these tanks or allow them in their jurisdiction. The reason is that it is difficult on site to truly verify that the tank meets the requirements of a design standard like UL 142 or UL 2085. Inspectors on assignment to review a new tank installation will find it very difficult to verify the grade of steel used, the dimensions and quality of welds, and the placement of structural components like stiffeners. The listing mark removes any doubt about such details.

Consider the following: A large brokerage firm purchased a diesel-powered generator to serve as its source of backup power. The design documents did not specify the selection of a listed aboveground storage tank. The owner's representative proceeded with the purchase and installation. The invoice for the tank indicated a purchase price of $3000. The contractor installed the 1000-gal base tank and generator set. Electricians worked for six days running electrical conduit and wires to the generator. When the installation was complete, the contractor called the fire department and requested an inspection. The fire official inspected the tank and could not find a nameplate with a listing mark. After reviewing the drawings and specifications it became apparent the tank was not listed. Once it was confirmed that the tank was not listed, the fire official issued a report to the owner requiring the removal of the tank. He also prohibited the tank from being filled with diesel. The owner and contractor were forced to remove the tank, disconnect the wiring, order a listed tank, wait seven weeks for delivery, and completely rewire the connections to the generator. Total cost of the error: $79,000. The cost of the listed tank—$4400.

This scenario illustrates that open dialogue between the tank owner and the fire official should begin as early as practical. Letting the local fire official know that a tank was installed on the day of the inspection can result in a expensive replacement and possible legal action by the authority having jurisdiction.

An even more frightening event is when a tank is installed and the installation is not in accordance with the adopted codes and standards. A fire official received a call from a business owner who reported that a 1000-gal gasoline storage tank had appeared next door at a vehicle storage yard. The business owner was concerned because of the tank's proximity to his building. The fire official could not find any record of an installation permit. When the fire official arrived at the business to investigate, he found a tank without an emergency vent or secondary containment. The tank was located on the property line. The electrical installation for the dispenser consisted of an extension cord connected to the dispenser electrical wires. In this case, the fire official condemned the tank, ordered the immediate removal of the gasoline and had the tank removed. The fire official also filed charges in municipal court, which cost the tank owner $1500 in fines.

2. Tank and Pipe Tests

Tanks and piping require a pressure test. Pressure tests are performed to determine if the assembled components are free of leaks.

NFPA 30 requires an integrity test of aboveground storage tanks with integral secondary containment. Secondary containment aboveground storage tanks require a test of the interstitial space using an air pressure or vacuum test. Secondary containment aboveground storage tanks are to be air tested at 3 to 5 psig, vacuum tested at 5.3 in. of mercury, or in accordance with the manufacturer's instructions.

The most common method of leak testing for piping systems is the hydrostatic test. Usually the test medium is water at ambient temperature. The model fire codes, NFPA standards, and ASME B31.3 require that the system be pressurized to 1.5 times the design pressure. The system must be held at a test pressure for at least 10 min, but can be reduced if the examination for leakage is complete.

Model fire codes and standards allow pneumatic testing of piping. Pneumatic tests may be performed in instances where water or other liquids are not acceptable, or where supports my not be adequate to carry the added weight of water. Pneumatic tests are potentially more dangerous than hydrostatic tests, and extreme care should be exercised [35]. ASME B31.3 requires the piping system be pneumatically tested to 1.1 times its design pressure.

The engineer may specify other types of pipe examinations. ASME B31.3 requires radiographic examination of a selected percentage of pipe welds. Visual examination of all flanged connections assembled with bolts is required. The designer is responsible for developing the pipe-testing procedures.

Piping is tested independently of the tank. Pressure testing a piping system connected to a tank can damage the tank. Consider a piping system designed to operate at 50 psig. ASME B31.3 requires a minimum hydrostatic test pressure of 75 psig. Conversely, the design pressure for atmospheric pressure storage tanks is 1 psig, but ASTs are never tested above 5 psig. Subjecting the tank to the piping's much higher operating pressure could cause the tank to catastrophically fail.

3. Component Inspection and Tests

Components require inspection and tests to ensure they are properly installed and functioning correctly. All of the tests should be performed in accordance with the manufacturers's instructions. Inspections and tests should be performed on:

Dispensers. Hose nozzles, hoses and dispensers should be listed for motor vehicle fueling. Remote pumps supplying dispensers should be listed for flammable-liquids service and have an operating pressure that does not exceed the pressure rating of the dispenser. When required, hoses should be equipped with breakaway devices.

Emergency shutoff switches, which should be properly identified and located in accordance with fire code and NEC requirements. When located outdoors, the switch and housing should be weather resistant.

Emergency shutoff valves, which should be properly anchored. Release of the fusible-link element should close the valve.

Emergency vents. Properly sized emergency vents should operate freely without binding. The operating pressure of the emergency vent must be greater than the normal vent and less than the tank design pressure. The discharge flow rate of the emergency vent must meet the flow rates specified on the tank nameplate.

Overfill prevention systems, which should be properly sized so it will stop the flow of liquid into the tank at a predetermined level. This is verified by measuring the length of overfill-prevention device and tank depth. The measurements should be compared to those specified by the manufacturer. Certain overfill prevention devices use a float valve attached to a lever. If the fill connection is located close to the tank shell, the float valve may not close because the lever strikes the tank shell. Inspection of the overfill prevention device may find markings that illustrate the proper orientation of the valve.

Leak detection systems. Leak detection probes should be tested to determine if they will operate when exposed to hydrocarbon liquids. When an alarm is required, its audibility and visibility—or both—should be inspected for adequacy.

A fire official should witness the component tests. Because of potential liability, fire officials must not operate or manipulate any of the controls or equipment. Jurisdictions have been sued because an inspector operated components or equipment during tests that later failed or did not perform as intended.

4. Electrical Inspections

An electrical inspector should examine the AST electrical system to verify the installation complies with NEC® and fire code requirements. Some essential inspection areas are:

Equipment: Electrical equipment located in hazardous atmospheres should be marked with the correct class, group, and division. If the equipment is in a Division 1 location, the equipment temperature code (T-code) must be lower than the liquid autoignition temperature.

Conductors: When installed at gasoline service stations or bulk plants, conductors should be listed as gasoline resistant. When conduit is used, it must meet NEC® requirements for Division 1 and 2 locations. Grounding should meet NEC® requirements.

Conduit seals: Should be located as required by the electrical code. Seals should be properly dammed and filled with sealant.

B. Maintenance Inspection

After a tank is approved for use and filled with product, inspections are necessary over the life of the installation. At a minimum, aboveground storage tanks need annual maintenance inspections. In certain cases, these inspections are required as a condition of having a fire department permit. Inspections help to ensure that the tank and its components are maintained in proper working order. Aboveground storage tanks use mechanical and electrical components. These components have a limited life and can fail. Maintenance inspections can help identify equipment deficiencies or improper modifications. Aboveground storage tanks also require inspection to determine their integrity. A tank that has been subjected to damage due to mechanical impact may be weakened or stressed to the point of failure (Fig. 29).

Figure 29 This aboveground storage tank was damaged when struck by a vehicle. The damage weakened the tank wall. Tanks in this condition should be drained of product and removed from service.

A maintenance inspection should evaluate the tank and equipment for:

Presence of leaks at hoses, connections, tank welds or around pump seals. One indicator of leaks is hydrocarbon stains around fittings and connections that are not normally made or broken.

The integrity of the containment. The containment drain valve should be closed.

The functionality of emergency shutoff switches and valves.

The operation of the normal and emergency vents.

REFERENCES

1. Department of Housing and Urban Development. Siting of HUD-Assisted Projects Near Hazardous Facilities: Acceptable Separation Distances from Explosive and Hazardous Substances, HUD-1060-CPD, Washington, DC, April 1987, p 51.
2. Siting of HUD-Assisted Projects Near Hazardous Facilities: Acceptable Separation Distances from Explosive and Hazardous Substances, p 15.
3. Steel Tank Institute. NFPA's Robert Benedetti responds to issues of code compliance and fire safety. Tank Talk, 1995, 10(2):1.
4. International Fire Code Institute. 1997 Uniform Fire Code, Vol. 1, Whittier, CA: International Fire Code Institute, 1997, p 118.

5. JJ Duggan, CH Gilmour, PF Fisher. Requirements for relief of overpressure in vessels exposed to fire. Trans ASME January, 1944, p 8.
6. National Fire Protection Association. Standard for Water Spray Fixed Systems for Fire Protection, NFPA 15, 1996. Section A-4.5.2. Quincy, MA: National Fire Protection Association.
7. National Fire Protection Association. Standard for Water Spray Fixed Systems for Fire Protection. NFPA 15, Section A-4.5.1.1 Quincy, MA: National Fire Protection Association.
8. National Fire Protection Association. Standard for Low-Expansion Foam Systems, NFPA Standard 11, Section 3-2.2. Quincy, MA: National Fire Protection Assocation.
9. Petroleum Equipment Institute. Recommended Practices for Installation of Aboveground Storage Systems for Motor Vehicle Fueling. Tulsa, OK: Petroleum Equipment Institute, 1992, pp 4–5.
10. International Fire Code Institute. Whittier, CA: Uniform Fire Code, Vol. 1, 1997 ed., p 24.
11. Office of Investigations and Analysis. Pollution Incidents In and Around U.S. Waters—A Spill/Release Compendium: 1969–1993. Washington DC: United States Coast Guard, March 1, 1998.
12. JL Goodier, RJ Siclari, PA Garrity. Spill Prevention and Fail-Safe Engineering for Petroleum and Related Products. Park Ridge, NJ: Noyes Data Corp., 1983, p 107.
13. Underwriters Laboratories Inc. Steel Aboveground Tanks for Flammable and Combustible Liquids, UL 142, 7th ed. Northbrook, IL: Underwriters Laboratories Inc., April 1, 1993, p 6.
14. Underwriters Laboratories Inc. Steel Aboveground Tanks for Flammable and Combustible Liquids, UL 142, 7th ed. Northbrook, IL: Underwriters Laboratories Inc., April 1, 1993, p 20.
15. American Petroleum Institute. Venting Atmospheric and Low-Pressure Storage Tanks—Nonrefrigerated and Refrigerated, API 2000, 4th ed. Washington, DC: American Petroleum Institute, September, 1992, Section 1.1.1.
16. Underwriters Laboratories Inc. Steel Aboveground Storage Tanks for Flammable and Combustible Liquids, UL 142, 7th ed. Northbrook, IL: Underwriters Laboratories Inc., April 1, 1993, p 20.
17. M Lankford. Devastating explosion in Texas relived. Tank Talk 9(4):1–2, 1994.
18. American Society for Testing and Materials. Standard Test Methods for Determining Effects of Large Hydrocarbon Pool Fires on Structural Members and Assemblies—Standard E-1529. Philadelphia, PA: American Society for Testing and Materials, 1993, p 11.
19. ASTM E-1529, p 12.
20. Underwriters Laboratories Inc. Steel Aboveground Storage Tanks for Flammable and Combustible Liquids—Standard 142. Northbrook, IL: Underwriters Laboratories Inc., 1996, p 19.
21. American Petroleum Institute. Venting Atmospheric and Low Pressure Storage Tanks, Nonrefrigerated and Refrigerated, API 2000, 4th ed. Washington: American Petroleum Institute, September 1992, p 8.
22. American Petroleum Institute. Overfill Protection for Storage Tanks at Petroleum Facilities—Recommended Practice 2350. Washington, DC: American Petroleum Institute, 1996, p 1.

23. JH Casiglia. Design bases. Piping Handbook, 6th ed. New York: McGraw-Hill, 1992, p B.57.
24. National Fire Protection Association, Flammable and Commbustible Liquid Code, NFPA 30, Sec. 2-4.3.1, 1996 Quincy MA: National Fire Protection Association.
25. RP Genereaux, CJB Mitchell, JB Charles, AC Hempstead, BF Curran. Transport and storage of fluids. In: Perry's Handbook of Chemical Engineering, 6th ed. New York: McGraw Hill Inc., 1984, pp 6–38.
26. American Society of Mechanical Engineers. Chemical Plant and Petroleum Refinery Piping, ASME B31.3, 1993 ed., p 5.
27. RC Getz. Chemical and refinery piping systems. In: Piping Handbook, 6th ed. New York: McGraw-Hill, 1992, p. c.288.
28. American Society of Mechanical Engineers. Chemical Plant and Petroleum Refinery Piping—ASME B31.3, 1993 ed., New York: American Society of Mechanical Engineers, 1993, Sec. 314.2.1.
29. American Society of Mechanical Engineers. Chemical Plant and Petroleum Refinery Piping—ASME B31.3, 1993 ed., New York: American Society of Mechanical Engineers, 1993, Sec. 323.4.2.
30. MW Earley, JV Sheehan, JM Caloggero. National Electrical Code Handbook, 6th ed. Quincy MA: National Fire Protection Association, 1996, p 519.
31. Underwriters Laboratories Inc. Power Operated Dispensers for Petroleum Products—UL 87, 10th ed. Northbrook IL: Underwriters Laboratories, October 24, 1995, p 4.
32. Underwriters Laboratories Inc. Valves for Flammable Liquids—UL 842, 8th ed. Northbrook IL: Underwriters Laboratories, April 30, 1997, p 1.
33. Underwriters Laboratories Inc. Valves for Flammable Liquids—UL 842, 8th ed. Northbrook IL: Underwriters Laboratories, April 30, 1997, p 2.
34. Underwriters Laboratories. Request for comments on proposed requirements for the proposed sixth edition of the Standard for Hose and Hose Assemblies for Dispensing Gasoline—UL 330, September 30, 1996, p A3.
35. EW Gerwin. Fabrication and installation of piping systems. In: Piping Handbook, 6th ed. New York: McGraw Hill Inc., 1996, p A.361.

23
Aboveground Storage Tank Installation

George H. Watkins
The Watkins Company, Fayetteville, Georgia

I. INTRODUCTION: "EXPERIENCE AS WELL AS SKILL"

The Petroleum Equipment Institute (PEI) in 1996 published a recommended practice that is designed to guide individuals in the installation of small-capacity (50,000 gal or less) aboveground storage tanks (ASTs). PEI/RP200-96, a document that forms a basis for much of what appears in this chapter, includes these words of wisdom:

> Installation of liquid motor fuel storage systems is a highly complex field, requiring a wide range of construction knowledge and experience. In addition to the proper design of an aboveground system, reliance on tank installers who possess both the experience and integrity to insist on doing the job right constitutes the greatest protection against ultimate tank-system failure and liability exposure. Written instructions alone will not convert an incompetent or under-supervised mechanic into a competent craftsman. The ability to recognize and react to unexpected, abnormal conditions encountered during a tank installation job requires experience, as well as skill [1].

II. GENERAL

In all cases, the manufacturer's installation instructions must be consulted before installing an aboveground tank system. The Petroleum Equipment Institute (PEI) and Steel Tank Institute (STI) have published reference information on correct AST installation procedures. PEI's RP200 covers the whole realm of AST installations, while STI's R912 and R931 thoroughly detail safe AST tightness-testing procedures.

III. LAYOUT

Ideally, by the time an installation reaches the layout stage, the owner has produced scale drawings and obtained necessary permits. If these things have been done, then the layout process simply requires locating the correct point on the property that is shown on the drawings, and orienting the system as shown. Some important considerations included in the layout process are:

- Distances from roads, property lines, and buildings
- Tank spacing requirements in multiple tank installations
- Traffic flow for both fuel delivery vehicles and others expected to utilize the fueling facility
- The location of utilities—both underground and overhead (Figs. 1, 2)

All of these factors should have been considered during the process of plan production and permitting. Nevertheless, a competent installer should be familiar with these requirements and prepared to recommend changes if these issues have not been dealt with properly prior to the layout process.

A. Separation Distances

Separation distance requirements may dictate whether a tank buyer purchases a traditional steel UL 142 tank, a fire-resistant tank, or a tank in a vault. (In fact, these limitations may well mean that an underground storage tank [UST] is the best or only solution.) Fire-resistant or vaulted tanks are generally allowed shorter separation distances by code, and therefore may be specified even if the application does not require insulated ASTs. Some codes further differentiate between private fueling operations and retail operations in the establishment of required separation distances—that is, less separation distance is allowed for private fueling facilities such as fleets [2].

IV. FOUNDATIONS

As a rule of thumb, undisturbed native soil will support the typical shop-fabricated aboveground tank. However, the repercussions of tank settlement can be severe, and the ability of the soil to support the tank and its contents must be considered. Probably the greatest threat posed by tank settlement is the resulting stress that may be placed upon the piping attached to the tank. This, of course, points out the

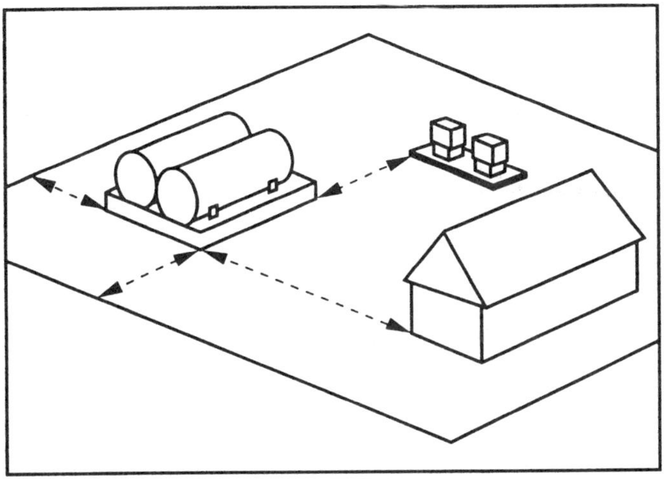

Figure 1 Spacing constraints.

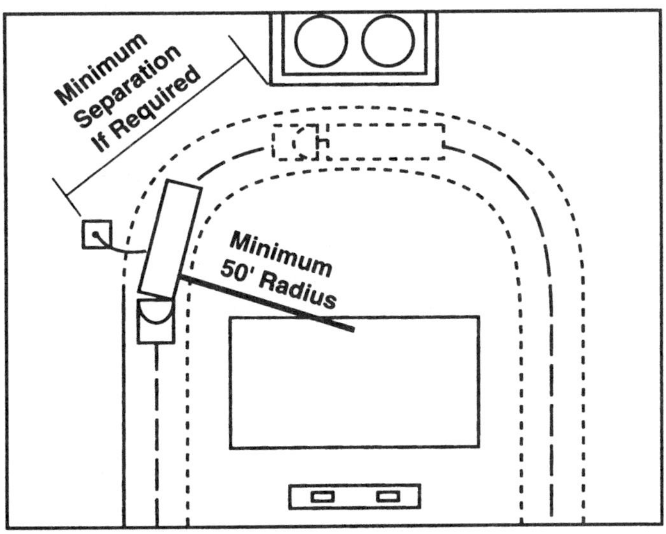

Figure 2 Good traffic flow.

importance of building flexibility into the piping system, as well as the need to prevent tank settlement. Even in areas where native soils unquestionably will support an aboveground tank, care must be taken to avoid locating a tank over a previously excavated area that may not have been properly compacted. If the soil support is suspect, several options are available to solve the problem with minimal costs:

Change the location of the tank to avoid an area with potential problems
Place the tank on a reinforced concrete pad designed to bridge across such areas
Replace the questionable soil with compactable materials

Under some conditions, a pad constructed of concrete is used. Stability is critical. Obviously, a larger tank resting on two saddles requires a foundation of greater strength than a small 500-gal tank resting on a full-length support or skid (Figs. 3–5).

In some areas of the country it may be necessary to have a professional engineer stamp the design drawings for the tank's support system. For example, a

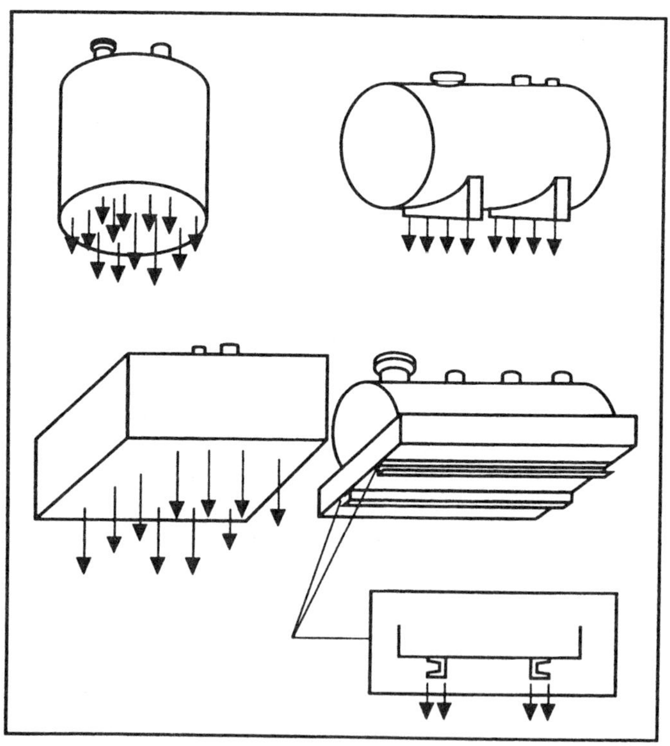

Figure 3 Load distribution points.

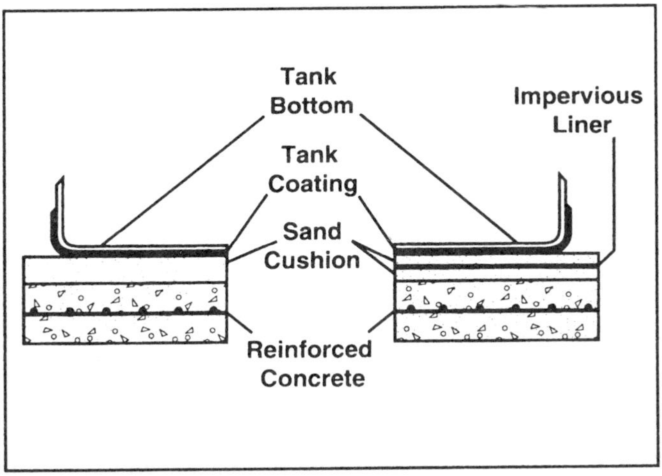

Figure 4 Vertical tank base section.

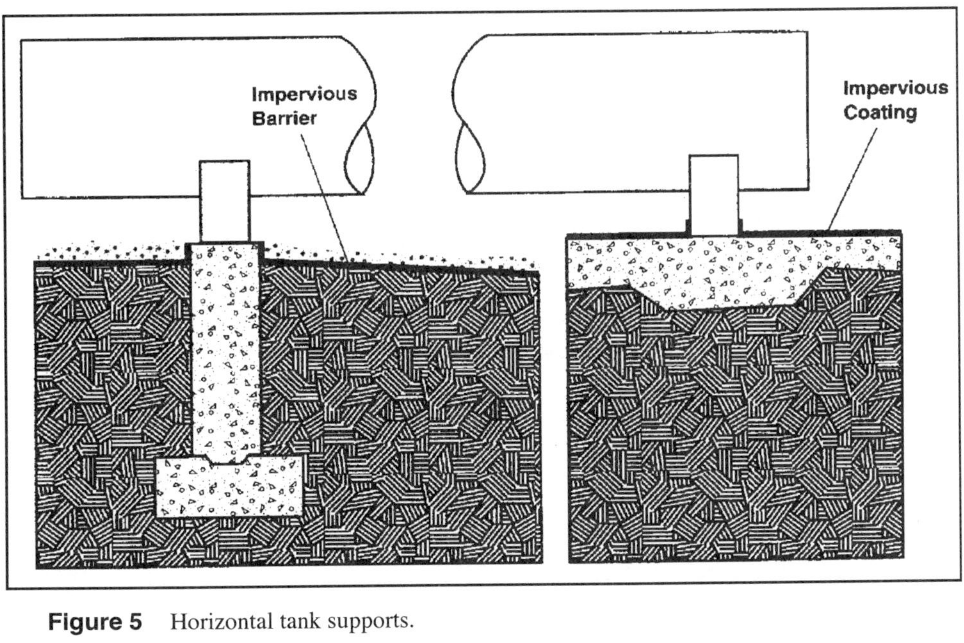

Figure 5 Horizontal tank supports.

portion of the West Coast is rated Seismic Zone 4, which means all aboveground storage tanks must meet minimum seismic requirements.

Tanks located in possible flood zones should be protected against flotation. This usually involves the use of straps that encircle the top of smaller tanks. These straps should then be bolted into a concrete pad, installed beneath the tank.

V. DIKES

If an aboveground tank installation involves the construction of a dike, several key points must be considered. First, although a dike that is not liquid tight may have value as far as firefighting is concerned, the environmental value of such a dike is questionable. Some dikes are constructed of slightly permeable soil. Some are made of concrete, a material that can crack over time (Fig. 6). There are codes and regulations that cover the capacity of dikes. These requirements vary from state to state and from municipality to municipality. Plans and specifications should address the capacity of a dike. But a competent installer must be aware of capacity requirements and prepared to deal with conflicts between the project plans and specifications and codes and regulations. A means of drainage must be provided for a dike, and the floor of the diked area must slope toward the drainage point. Drainage control should be outside the dike and accessible in the event of a fire.

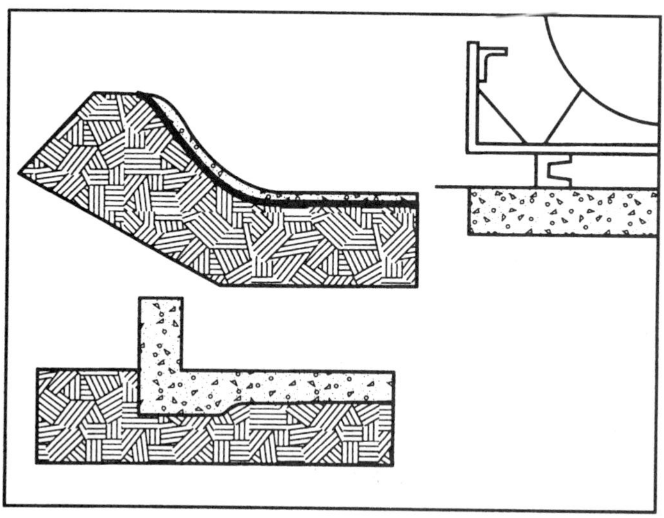

Figure 6 Typical dike sections.

VI. TANKS

Aboveground storage tanks come in all shapes and sizes. A tank may be designed for vertical or horizontal operation. A tank may be one of many configurations from single-wall steel to double-wall encased in concrete. The tank may include a shop-fabricated dike—either attached to the tank or as a separate item (Fig. 7). The installer must provide equipment adequate to lift and move the tank whatever the size, weight, or configuration. He must also inspect the tank to be certain that it has not been damaged in shipment or unloading. He should always refer to all information provided by the tank manufacturer, and follow any installation instructions from the manufacturer.

VII. CORROSION PROTECTION

Horizontal tanks are usually placed on supports, either saddles or full-length supports, so corrosion of the primary tank is not a concern. However, vertical ASTs that are not elevated—that is, the tank bottom sits directly on the ground—must incorporate corrosion protection. Depending on the square footage of the steel exposed, a galvanic anode system may work well. Larger surface areas, or installations where the AST is electrically connected to a steel underground piping sys-

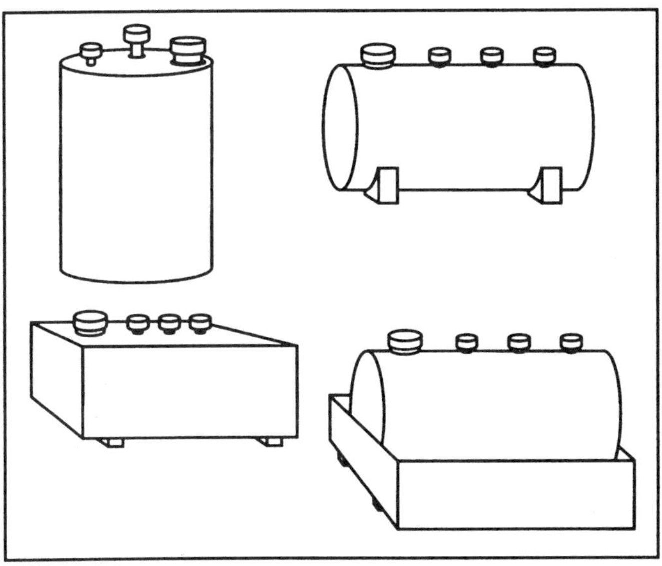

Figure 7 Typical tank profiles.

tem, may require an impressed current system. Such systems are commonly used where the bottom of an existing tank is not coated (new vertical tank bottoms should be coated). In such cases, a corrosion specialist should be contracted to design an impressed current system. The National Association of Corrosion Engineers (NACE) and STI both have specific written standards addressing AST tank bottom corrosion protection. One method of minimizing corrosion is to design the foundation to drain surface water away from the tank.

VIII. GROUNDING

For tank installations without cathodic protection, a static electricity grounding system should be installed on the tank in accordance with applicable electrical and fire code standards. For tanks with cathodic protection, additional grounding may not be needed. However, NACE standards should be consulted to provide the tank with appropriate protection from static electricity without disruption of the corrosion protection system.

IX. ACCESSORIES

Accessories include components that, together with the tank, will create a system to store and dispense liquid. Plans and specifications should clearly spell out the requirements for pumps, valves, fills, vents, and other accessories. Many installations are done without adequate detail on plans and specifications. This leaves many important decisions up to the installer as to what accessories to use, how to use them, and where to use them. Even when detailed plans and specifications are provided, the installer may discover that the design reflects a lack of familiarity with equipment requirements for aboveground tank systems. Experience has shown that critical mistakes are frequently made by failing to include, or improperly placing, vital components. The following paragraphs describe several typical fuel systems and some accessories necessary for each.

There are several typical methods for dispensing fuel from aboveground tanks. The simplest method for vehicle fueling is to mount a suction-type pump directly into a bung at the end of a tank (Fig. 8). This system obviously is not suited for larger-diameter tanks. A larger-diameter tank would elevate the pump to a height that would not be easily accessible to a person attempting to refuel a vehicle. This simple system requires an antisiphon valve to prevent product from siphoning through the pump. This could occur if the hose were to become defective below the tank's liquid level, or if a vehicle pulled off the nozzle. Some manufacturers build antisiphon devices into pumps of this type, and therefore, an installer should determine if such a device is required.

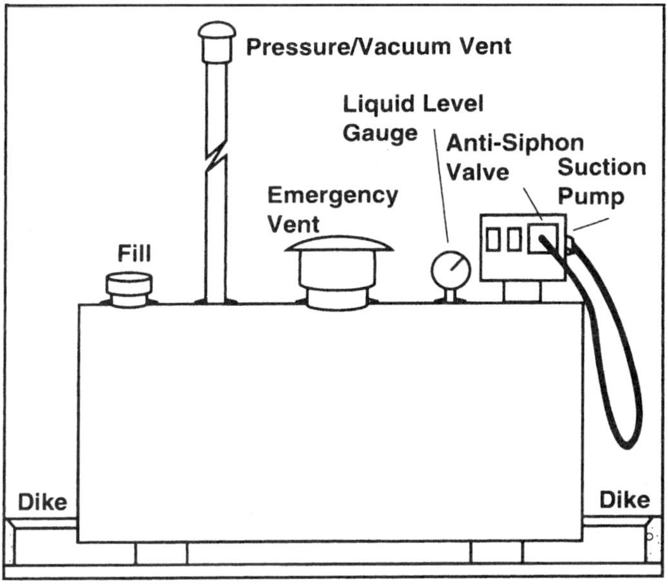

Figure 8 Tank-mounted suction system.

Other accessories are the fill connection, which may be located on top of a small-diameter tank, within easy reach of fuel delivery personnel. Installers must be aware that aboveground tanks are normally filled by pumping the product from the delivery truck into the tank, and that overfill equipment designed for underground tanks, which are gravity filled, may not function properly under these conditions.

All tanks must have a properly sized atmospheric vent located at a height and distance from buildings consistent with prevailing codes. All aboveground tanks must have a properly sized emergency vent. Firefighters have been killed because aboveground tanks were installed without emergency venting. The emergency vent is designed to relieve excess pressure that builds up in a tank during a fire, thus preventing the vessel from exploding. Double-wall tanks require an emergency vent for both the primary tank and the interstice.

Another common method for dispensing fuel into vehicles from aboveground tanks is to use a conventional suction pump mounted adjacent to or remote from the tank, usually below the tank's liquid level (Fig. 9). Additional equipment will be required with this arrangement, beyond that required for the previously described small-diameter tank with a pump mounted on top of the tank.

First, this installation allows the use of larger-diameter tanks. With larger-diameter tanks, steps or ladders may be required to gain access to the top of the tank for filling and gauging. The need for steps or ladders may be avoided by mak-

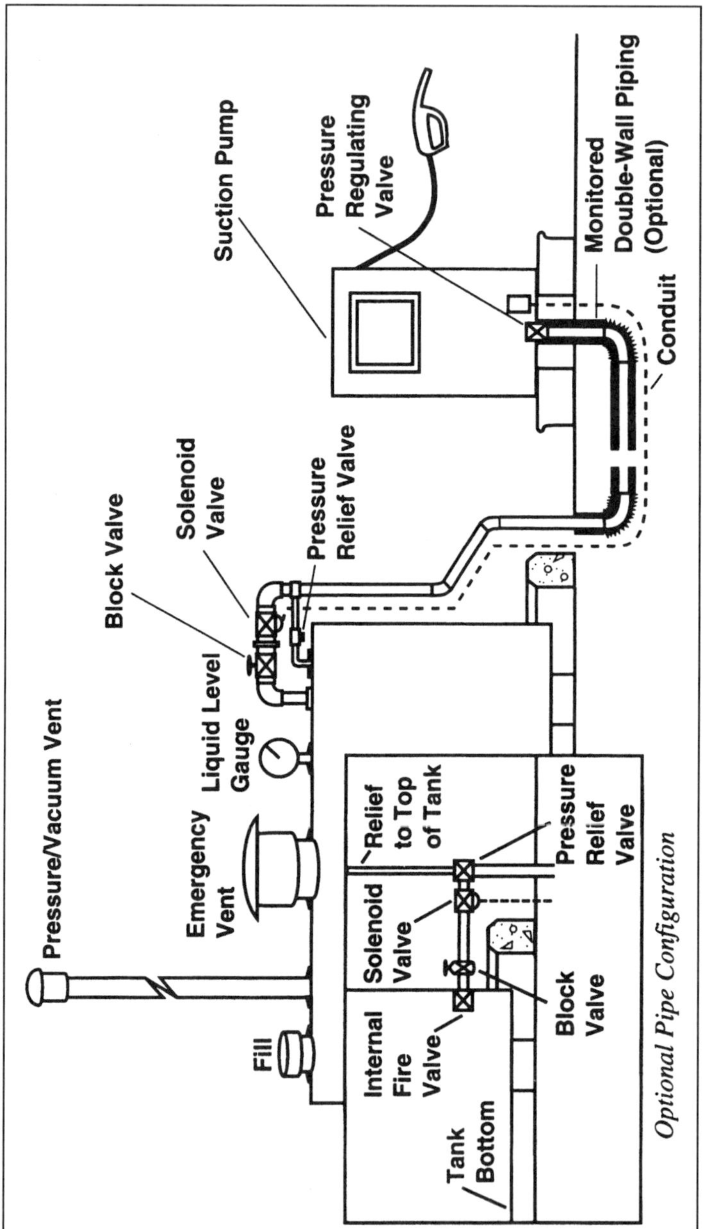

Figure 9 Suction system.

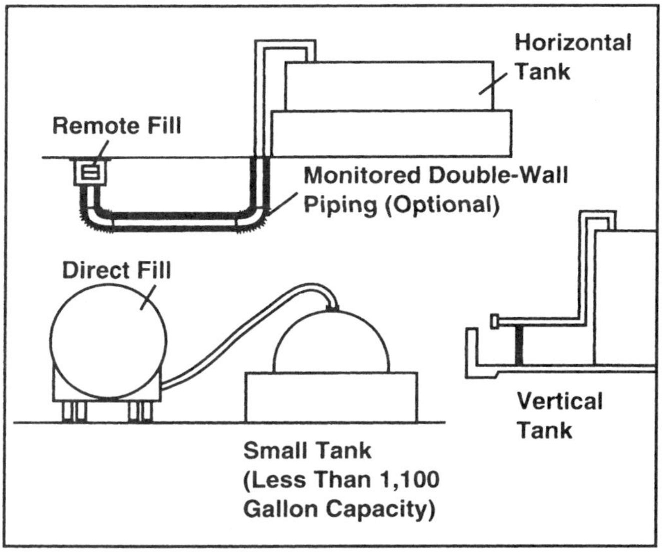

Figure 10 Typical fill methods.

ing the fill accessible and by providing a tank gage. The fill may be piped down to a more convenient location either next to the tank or remote from the tank, and a visible gauge may be mounted on or near the tank. In instances where the fill connection is piped to a level lower than the top of the tank, a check valve or a dry-break connection must be incorporated into the line to prevent backflow (Figs. 10, 11).

Many systems such as this utilize underground piping between the tank and the suction pump. Any piping located below the tank's liquid level has the potential to leak continuously unless antisiphon devices are incorporated into the piping system. There are two antisiphon devices used most often in aboveground tank installations. One is a solenoid valve wired to open when the pump is running. The other is a check valve designed with a spring, which is set to hold the valve seated unless the pump is running. A block valve should be located at the point where the suction line connects to the top of the tank to manually block the line. In cases where the piping exits the tank below the liquid level—and in areas where codes permit such an installation—an approved fire valve should be used as the first fitting out of the tank. A fire valve is a device designed to close automatically when subjected to intense heat. A fusible link will melt during a fire causing a spring-loaded poppet to close, cutting off the flow of liquid (Fig. 12).

Of great importance is the use of a relief valve in any segment of piping that can be blocked at both ends. A blocked section of piping exposed to the sun for a short period of time has the potential of reaching pressures far exceeding the pipe system's design criteria. Such excessive pressures can cause piping to leak or rup-

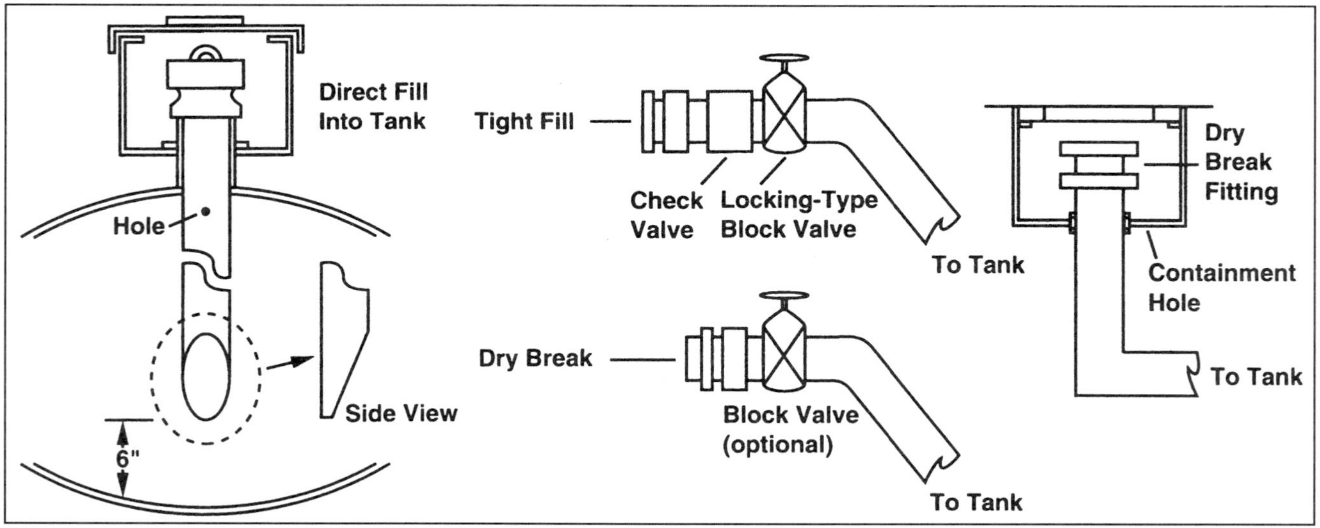

Figure 11 Typical fill connections.

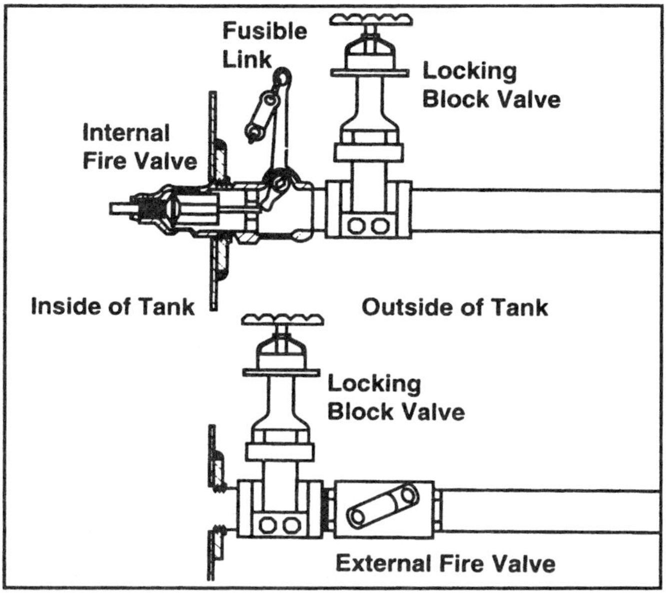

Figure 12 Fire valves.

ture. One additional item is necessary with a system in which a suction pump is used below the liquid level in the tank—a pressure-regulating valve. This device is located immediately below the pump and is designed to allow product to flow only when the pump is running. It also is designed to break in a manner that stops the flow of product if the pump is knocked over (Fig. 13).

A third common method for dispensing fuel into vehicles from an aboveground tank is by utilizing a remote pump at the tank and a dispenser at the fueling point. One advantage to this design is that higher-speed fuel delivery can be achieved. Another advantage is that greater distances between the tank and the fueling point are possible. This system has the same requirements for fills and vents as those previously covered. This system design may utilize any of several types of pumps, but the most commonly used is the standard underground storage system's submerged pump. Since using a remote pump creates pressurized piping and because a severe loss of product can result from such a release, a leak detection method must be part of the piping system. If the piping is underground, any of the many leak detection systems developed for underground tank systems can be used effectively. The requirement for an antisiphon device exists with this system, as product can siphon through a submerged pump when it is not running.

A block valve should be used at the pump outlet if a submerged pump is used, and the need for a pressure relief valve remains. A conventional industrial pump

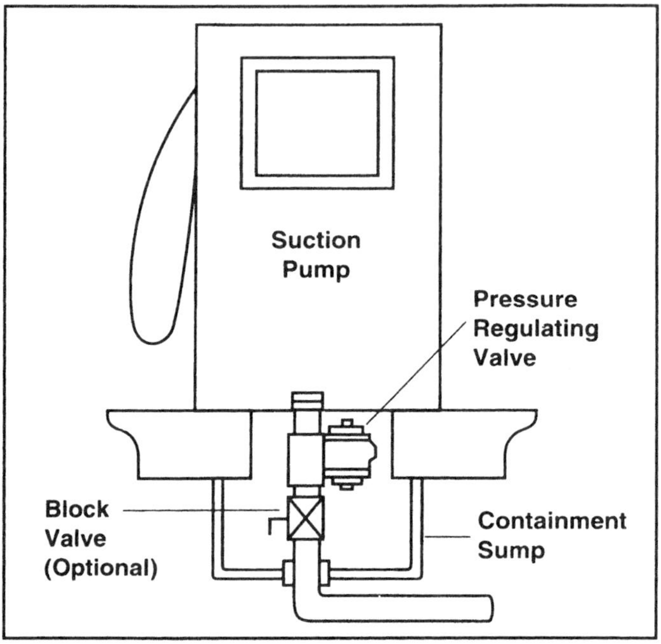

Figure 13 Pressure regulating valve.

could also be used, mounted at the base of the tank below the liquid level, with pressurized piping connecting to a dispenser. This system would actually incorporate both the pressure piping between the pump and the dispenser and suction-type piping between the tank and the pump. Under this scenario, block valves should be placed at the tank outlet, at the pump inlet, and at the pump outlet, with pressure relief between all these points. Flanges (or unions) should be used to facilitate maintenance on the pump, and should be arranged such that these fittings are located between the pump and the valves. This will allow the valves to isolate the pump, and the flanges to be disconnected for pump removal. A remote pumping system requires that a fire/impact valve be used beneath the dispenser.

A fourth type of system seen frequently uses an aboveground tank to store fuel for a boiler or a standby generator. Many of these systems require that the piping terminate inside a building, causing critical fire safety concerns. Typically, methods used to transfer the fuel from the tank to the boiler/generator, for the most part, are similar to those used with vehicle-fueling systems, using either suction-type pumps at the boiler/generator or remote pumps at the tank. Requirements for fills, vents, block valves, antisiphon valves, and relief valves are also similar to vehicle fueling.

If a remote pump is used, the need for leak detection on pressure piping remains as important as with other systems discussed. These systems do not use standard vehicle-fueling pumps or dispensers. So, pressure-regulating valves and fire/impact valves designed for use with vehicle-fueling systems are not normally used with boiler/generator systems. If the tank is located at an elevation higher than the boiler/generator, the possibility exists that fuel could siphon into a building, causing the requirement for antisiphon devices to be even more critical.

One thing that differentiates a boiler/generator system from a vehicle-fueling system is the product return line, which transports excess fuel back into the tank. A return line is not common in vehicle-fueling systems. Return lines are typically one pipe size larger than the supply line. Unlike the supply line, which by necessity terminates below the liquid level (within 6 in. of the tank bottom), the return line should terminate inside at the top of the tank, above the liquid level. By not allowing this line to terminate below the liquid level, any potential siphon problems from the tank through the return line are prevented unless the tank was overfilled.

A return line may use a pump at the boiler/generator to pump this excess fuel back to the tank. In this case, an additional pressure line is created, with its accompanying need for leak detection. One popular method of providing leak detection for both the supply and return line is to place both lines within a common containment pipe, with liquid sensors at the low point in the piping system. According to NFPA 31, a standard of the National Fire Protection Association, the use of valves in the return line should be avoided. A valve left closed in a return line could cause fuel to back up and overflow into a building. When an underground tank that stores fuel for a boiler/generator system is replaced with an aboveground tank, careful consideration must be given to the existing return line. The return line may have worked fine by gravity into the underground tank, but a completely different return system may be required for the AST.

X. PIPING

Reference has previously been made to the need for flexibility in a piping system. Also, in the preceding sections covering various systems, there are several references to proper piping methods. As a general rule, any underground piping used in conjunction with aboveground tanks should follow the requirements for piping associated with underground tanks. These requirements cover corrosion protection as well as leak detection. Aboveground piping should be made up of Schedule 40 steel pipe and 150# steel or nodular iron fittings. Such aboveground piping should be protected from corrosion with paint. Flanges or unions should be used at convenient points for future needs to disconnect. The transition from aboveground piping to underground piping should be accomplished in such a manner that avoids placing steel piping unprotected from corrosion in contact with the soil. Underground piping also should not be exposed aboveground, if it is not designed for aboveground use. This transition is best accomplished within a containment sump, with the actual connection preferably using a fire-rated flexible connector.

As with USTs, a predominant cause of leaks in an aboveground tank system is piping problems. Any underground piping should meet the requirements of the EPA regulations 40 CFR Part 280, which requires that any underground piping be protected against corrosion or constructed of corrosion-resistant materials. Piping, valves and fittings located aboveground must be suitably protected against vehicular impact through the use of guard posts or other approved means.

XI. PRECONFIGURED SYSTEMS

Underwriters Laboratories (UL) in 1997 published the UL 2244 standard (Outline of Investigation for Aboveground Flammable Liquid Tank Systems For Motor Vehicle Fuel Dispensing), based on requests from inspectors who evaluate complete tank systems. The UL 2244 approach simplifies the approval process for the authority having jurisdiction (AHJ). At the time of installation, an AHJ can simply determine if a 2244 system complies with code by checking the Code Compliance Verification List developed by UL. This list must be shipped with each 2244 tank system.

All equipment that is a part of the system has been evaluated by UL as part of an individual listing. To qualify for use on a 2244 system all of the equipment must be evaluated by UL. Some of the equipment, such as the tank, pump and emergency vents, for example, must be UL listed. Other accessories, such as a normally open vent or block valve, do not require UL listings [3].

XII. TESTING

Prior to the introduction of liquid, the entire system should be tested for leaks. The tank should be tested by following the manufacturer's recommended procedure, which may or may not involve air pressure and soap.

While it is not often that a leak is found, an on-site tightness test is required to ensure that no damage has occurred in shipping and handling. Single-wall tanks should be pressure-tested with air, up to the manufacturer's maximum recommended pressure. Horizontal tanks are usually tested between 3 and 5 psig whereas vertical tanks are usually tested between 1 ½ and 2 ½ psig. A soap test is conducted simultaneously, since it may be impossible to detect a slow leak simply by looking at a gage. For example, a leak from a weld seam would be quite obvious from the small stream of bubbles.

Double-wall ASTs are tested a bit differently. The primary tank is pressurized first; then the air is transferred from the primary tank to the interstice (the space between the two tanks). This prevents the interstice from being overpressurized. It is also quite common to tightness test a double-wall tank with vacuum, where the tank arrives at the job site with vacuum on the interstitial space and a gauge on the monitoring pipe indicates the vacuum level.

Steel piping should be air and soap tested at 50 psi. All piping connections should be sprayed with a soap and water solution, observing for bubbles. Nonmetallic piping should be tested according to the manufacturer's instructions.

Many local fire inspectors or building code officials require that a representative from their offices observe these tests. A prudent installer should also request that a representative of the owner observe these tests. Obviously, the installer should initially perform all tests and repair any leaks prior to scheduling local officials or the owner to witness the final test.

XIII. DOCUMENTATION

With any tank installation, it's a good idea to carefully document the event. Photographs, red-lined drawings, bid specifications, warranty information and a list of contractors involved in the installation should be kept in a safe location. After all, the tank owner is responsible for any future site contamination that may result from improper installation procedures. Thorough documentation of such procedures is simply good business sense. Of course, it's also prudent to make sure the installing contractors are reputable and experienced in AST installation. Many state petroleum or contractor associations conduct AST installation training and testing for installers. A quick phone call to the regional association is well worthwhile [2].

REFERENCES

1. Petroleum Equipment Institute. RP 200-96, Recommended Practices for Installation of Aboveground Storage Systems for Motor Vehicle Fueling, 1996.
2. L Grainawi. Aboveground storage: Part I—Decoding shop-Built tanks. Chem Eng. August:98–103, 1996.
3. Steel Tank Institute. UL Issues 1st Proposed Edition of New UL 2244 Systems Listing. Tank Talk XIII(3), 1998.

24

How to Specify AST System Equipment

Charlie Glab
Morrison Brothers Company, Dubuque, Iowa

The two main questions when specifying AST equipment are:

What is required?
What is desired?

Requirements involve physics and laws. Physics covers system function, fluid dynamics, and other physical properties. Laws cover regulations, codes and standards imposed by federal and state policies, or local ordinance.

What is desired involves issues beyond basic requirements such as aesthetics, efficiency and maintenance factors, degree of automation, and return on investment. It is what you want to get out of the system beyond simply meeting requirements.

Optimizing goes beyond minimum requirements. Regarding equipment, many products are alike in basic function. It is only when the specifier considers issues beyond the basics that unique features and benefits become more evident. Some people have taken years to develop equipment specifications. Much has evolved through trial and error—and a certain effort must be made to keep up with regulations. There may be no perfect specification, but there are plenty of successful ones. The most successful specs have developed from a clear understanding of requirements and objectives.

The following is an outline identifying equipment requirements, applications, and options for equipment used on shop-fabricated, aboveground, atmospheric tanks that store flammable liquids. These are tanks that are less than 50,000 -gal capacity and have a maximum operating pressure of 1.0 psi.

I. BASIC PHYSICAL REQUIREMENTS

Study the basic physics of the system and understand the requirements. The tank provides the storage. The transfer network provides the means of moving product in and out of the tank. The venting network allows the product transfer and storage to occur under proper and safe conditions.

II. TRANSFER NETWORK

The transfer network involves filling and dispensing. Filling can be handled on top of each tank through a fill port, or it can be handled via piping from a remote port, or via piping from another tank. Dispensing occurs via piping to a dispenser, meter, loading rack, or to another tank. Fill lines and dispensing lines may be shared or separate. Most commonly, in a single small-tank system, the functions are separate. In a large-tank system with multiple units, these functions may share piping. In a manifolded system, they share the network by design.

A. Fill Port

1. Top Loose Seal

On the simple side, common to small utility tanks, the fill port is located on top of the tank and consists of a riser pipe with a cover (Fig. 1). The cover is commonly a 2-in. aluminum or iron fitting with male or female threaded adapter and lockable hinged cover. Normally it is not a tight seal and certain styles actually function as a vent/cover combination.

2. Top Tight Seal

A top-fill riser pipe with a tight seal is becoming more common on small tanks (Fig. 2). This cover is removable and is either threaded to the adapter or features a cam-and-groove connection. The most common size on small tanks is 2-in. The cam-and-groove design is becoming more prevalent because of the need for tight-fill.

3. Top-Fill Spill Containment

The fill port is becoming less simple due to spill containment and overfill prevention. Top-fill spill containment is handled by means of a sump or bucket either welded to the tank or installed as a separate piece of equipment that is threaded to the riser pipe (Fig. 3). Typically a 4-in. connection, the capacity of the detachable bucket ranges from 3 to 7 gal and it may come with an internal drain valve. These containers usually have a hinged cover that will not seal tight to allow for ventilation.

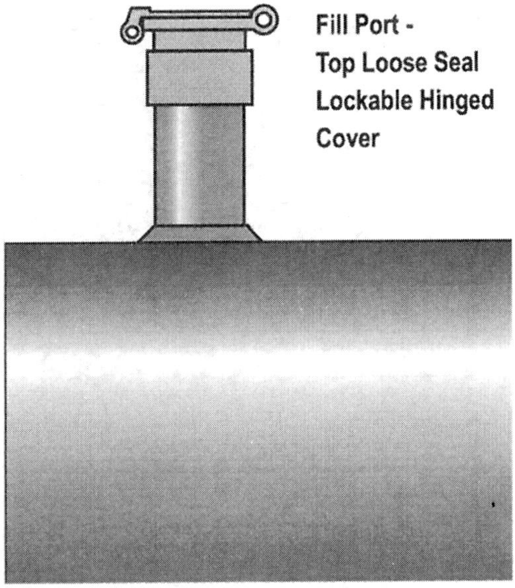

Figure 1 Top fill-hinged cover.

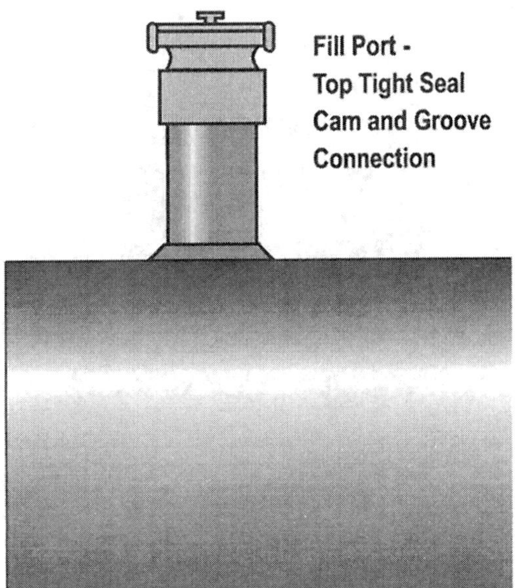

Figure 2 Top fill-tight seal cover.

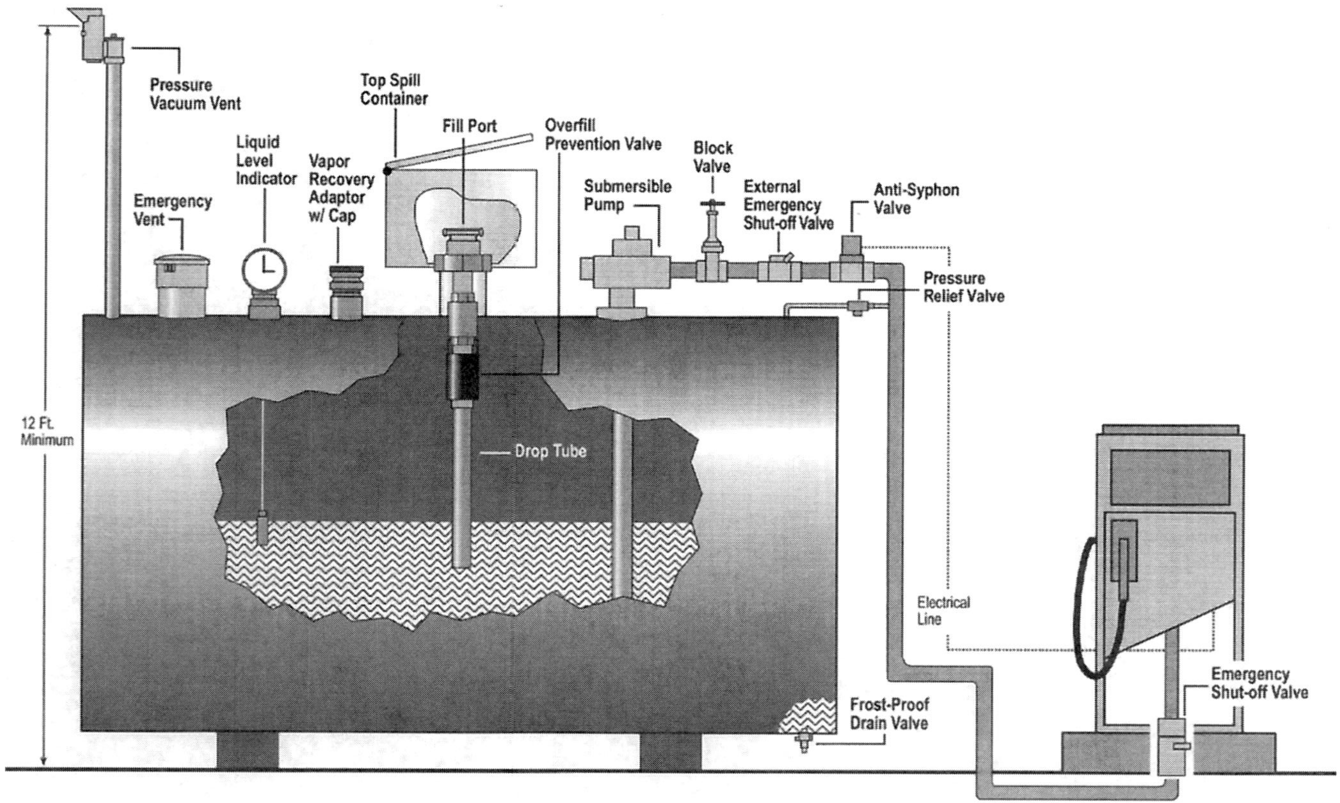

Figure 3 Aboveground fuel storage-pressure system. Horizontal cylindrical tank with top fill and remote dispenser.

4. Top Overfill Prevention

Overfill prevention is handled by means of a valve (Fig. 3) that actually replaces the simple adapter in the fill line. These valves are designed with a float mechanism that activates an automatic shut-off at some preset level (typically 90 percent full) in the tank. Sizes of these valves are most commonly 2 in.—with 3 in. and even 4 in. available. (Note that the size of the valve normally requires a larger tank opening, i.e., 2 in. requires 4 in., and 3 in. requires 6 in.) These valves are used in combination with a drop tube—and most likely a spill container—and can be configured in a number of ways. The specifier is seeing more engineered options today than any other time in AST history. This makes it even more critical to study the equipment thoroughly before making any final decisions.

5. Remote

Remote fill ports (Fig. 4) are common for larger tank systems simply due to restrictions on access to the top of the tank. They are also used to consolidate lines in a fill dock area on a multitank system. A variety of remote port configurations are used. The line from the fill opening to the tank is commonly constructed of rigid piping that may follow grade to the tank where it is piped up and over to the top. An overfill prevention valve may be installed in the line on top of the tank. A spill containment basin may be installed at the point of fill. These are either custom fabricated or tank owners can choose from several commercially available designs. Typically the line connects to the back of the unit and terminates with a cam-and-groove fitting mounted horizontally inside the basin. A hinged cover in front provides access to the fitting.

6. What to Do with Product After Overfill Shutoff

One issue with tight-fill and overfill prevention involves product left in the line when a shutoff has occurred. Often, the means to drain the delivery hose when overfill shutoff occurs is through a system design engineered especially for that purpose.

B. Pipeline Equipment

On small stand-alone tanks for motor fueling, dispensing product out of the tank usually involves a pipeline connected to a dispenser or pump (Fig. 4). In many cases the dispensing device is mounted on, or adjacent to, the tank. In this pipeline there may be a block valve, check valve, antisiphon valve, emergency shutoff valve, line strainer, and provision for pressure relief.

1. Block Valve

This device allows you to shut off the flow of product in the pipeline. It may be known by more common names such as a gate valve or ball valve. It is used to iso-

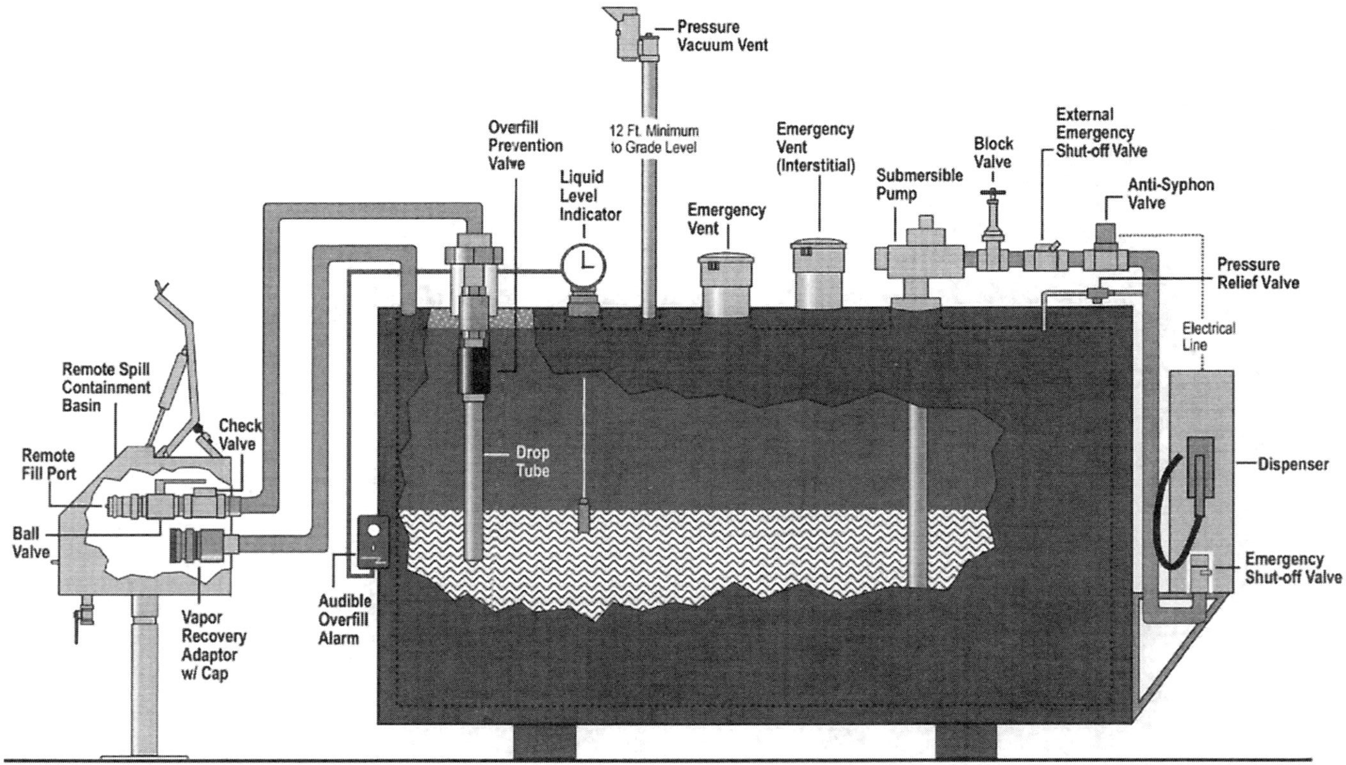

Figure 4 Aboveground fuel storage-pressure system. Rectangular double-wall tank with remote fill and side mounted dispenser.

late a section of the line for operational or maintenance purposes. These valves are usually brass or ductile iron. They come in a variety of sizes, threaded or flanged, and can be specified as lockable and with provision for pressure relief.

2. Check Valve

Installed in the pipeline, a check valve allows product to flow in one direction only. It contains a poppet that either lifts up or swings out of the way as product flows in one direction, but seals against the seat, preventing product from flowing back in the other direction. It is used in a variety of applications from maintaining prime on a pump to preventing backflow in a manifolded system. These valves are most commonly brass or ductile iron, and equipped with either metal-to-metal or soft seat. They are available in a variety of sizes—threaded or flanged—and can be specified with provision for pressure relief.

3. Antisiphon Valve

An antisiphon device is used to prevent product from leaking out of a tank if a break in the line occurs downstream below the liquid level of the tank. The two main varieties used for this purpose include an electronic solenoid valve and a mechanical spring-loaded valve. The electronic solenoid valve is normally closed and opens only when activated by a signal such as when a pump is turned on. The mechanical variety employs a spring set heavy enough to counteract the maximum head pressure of the tank. Often this setting is high enough to create flow restriction and problems, especially on suction systems.

4. Emergency Shutoff Valve

An emergency shutoff valve shuts down the flow of product when either a fire or impact occurs on the piping or tank. They are available in three main configurations. An internal style, external style and under-the-dispenser style.

a. Internal Style. The internal style (Fig. 5) is used most commonly on large vertical tanks and connects to a tank bung and piping at the base of the tank. The poppet of the internal style is located inside the tank.

b. External Style. The external style (Fig. 5) is a valve that can be mounted anywhere on the pipeline, usually as close to the tank as possible, and is most commonly used on tanks with piping that connects to the top of the tank. The external and internal types are both available in brass, ductile iron, and steel, threaded or flanged. The internal style also may come with a lockable option. It should be noted that the lock is only to be used for security purposes when the valve is closed.

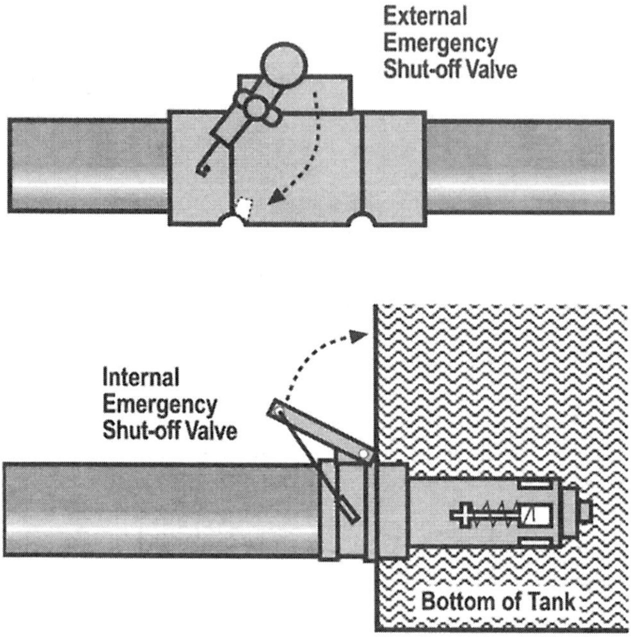

Figure 5 Emergency shut-off valves.

c. Under-the-Dispenser Style. The under-the-dispenser style (Fig. 3) is the same one used for underground storage tanks (USTs) and is commonly called an impact valve. For AST systems the application is virtually the same as USTs, which means the mounting must be rigid and placed directly under the base of the dispenser. These valves are left open in operation and are designed to shut off flow in case of fire or impact.

5. Line Strainer

A line strainer is often used in a piping system by placing it in the flow path immediately in front of a pump, meter, or other sensitive equipment. For various reasons, debris such as nuts, washers, or other articles can enter the line and damage equipment if allowed through. A line strainer can prevent costly damage, although the strainer basket will need to be cleaned out periodically to maintain proper flow rate performance. Line strainers come in a variety of configurations and a variety of basket screen mesh sizes.

6. Pressure Relief

Pressure relief is a provision (Fig. 3) that controls the maximum pressure in a pipeline. In AST systems, piping is commonly installed aboveground, and thus exposed to the sun. Also, depending on the design, product can be isolated such as between two closed valves. When this condition exists, overpressurization can occur in the line when the sun heats up the pipe. If no provision for pressure relief (also called expansion relief) exists, product in the line will seek the weakest point, such as valve packing or a gasket area, and leak out. Pressure relief is available as an optional feature on certain valves or it can be designed into the system using a small pressure relief fitting. The function allows pressure and a small amount of product to flow back to the tank. This limits the amount of pressure that can build up in the line to well below the working pressure of the system.

III. VENTING NETWORK

Accommodation for venting (Fig. 3) involves "normal venting" and "emergency venting." Normal venting is the means by which the tank "breathes" during normal operation such as filling and dispensing. During filling, pressure builds up and the tank must "breathe out." During dispensing, a vacuum is created, which forces the tank to "breathe in." For an AST exposed to the sun, when the day heats up, pressure builds up and the tank needs to breathe out. When the sun goes down and the tank cools off, a vacuum can be created and it needs to breathe in again. Accommodation for normal venting is a basic physical requirement for any AST system covered in this section.

A. Normal Vent

1. Open Style

There are various types of normal vents. The main distinction is open vent style (Fig. 6) versus pressure vacuum style. The open vent is simply an orifice typically designed for mounting on a pipe with a rain cover and screen. It allows air in and out of the tank without any limit beyond flow capacity. The open vent is normally used with fuels that have lower evaporation rates, such as heating oil, because air quality concerns and loss of product will be less compared to fuels with high evaporation rates such as gasoline.

2. Pressure Vacuum Style

Pressure vacuum vents (Fig. 3) also allow air into and out of a tank but under certain limits referred to as pressure and vacuum settings. These valves are built with a poppeted mechanism that shuts off air flow up to the point of the setting. Once the tank reaches that limit, the vent opens and allows air to flow. Normal settings for these vents are 8 oz/in.^{2} pressure and 1 oz/in.^{2} vacuum. These vents are pre-

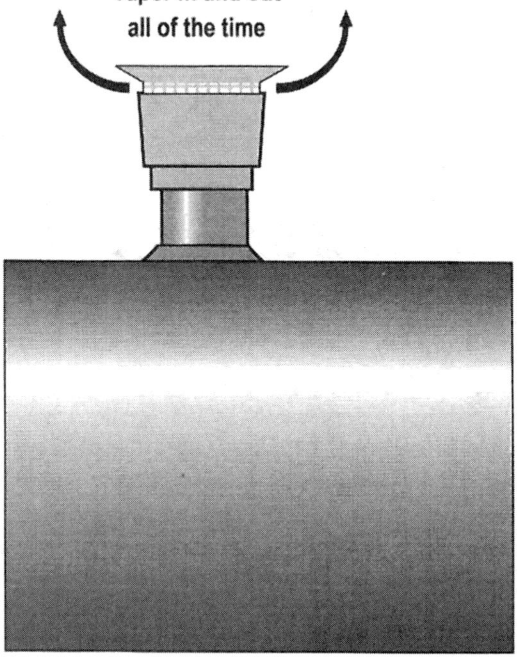

Figure 6 Normal vent-open vent style.

ferred and often required for use when storing liquids with high evaporation rates such as gasoline. A pressure vacuum vent controls product loss by limiting the opportunity for evaporation.

3. Design Considerations

Certain fire codes require that vents exhaust in an outward and upward direction. Most often they are required to be located at least 12 ft above grade and the exhaust should be directed away from any buildings or areas of activity such as transport unloading areas. Vent sizing is important. The normal vent must be able to handle the volumes generated by the liquid flow rates. A rule of thumb is that the vent size should match the liquid transfer piping size.

One caution is that there are distinct pressure vacuum vents designed for AST and UST applications. The UST vents have less capacity; however, they occasionally can be included on AST motor-fueling systems with low flow rates. The best process for choosing a normal vent includes a review of important criteria that govern flow rates and venting capacity.

B. Emergency Vent

Emergency venting (Fig. 3) is the means of relieving dramatic pressure built up in a tank due to overpressurization such as in an exposure fire. During a fire, the exposed tank acts like a tea kettle. The product inside heats up and pressurizes the tank at an accelerated rate. The emergency vent, if properly specified and functioning, will exhaust this pressure from the tank at a rate that keeps the tank below the operating pressure limit. Operating within the pressure limits, even during a fire, will pose no threat for structural damage, or explosion throughout the term of the emergency. Although not required for basic fluid dynamics, emergency venting is the single most important aspect for this type of AST system to ensure safe operation.

1. Options

Emergency venting can be handled by a special tank design called "weak roof-to-shell." This practice is expensive and makes for more difficult handling in transport and installation of the tank. Today, the most common practice is to install a venting device on the tank. This device is either a product made specifically for the purpose, or an adapted tank feature such as a loose-bolt manhole. The most common types of vents are fittings designed as a vent; these come in either the pop-up, weighted-cover style or the flip-open, hinged-cover style.

2. Flip-Open Cover Style

The spring-loaded, hinged cover (Fig. 7) is normally set closed and has either a single-stage or two-stage setting release. When the pressure setting is reached, the vent activates and springs open. The vent does not reset itself.

3. Pop-Up Weighted Cover Style

The pop-up style, which is the most common method used today, simply employs a weighted cover. When activated, it rises off the seat then lowers and resets itself when the pressure recedes.

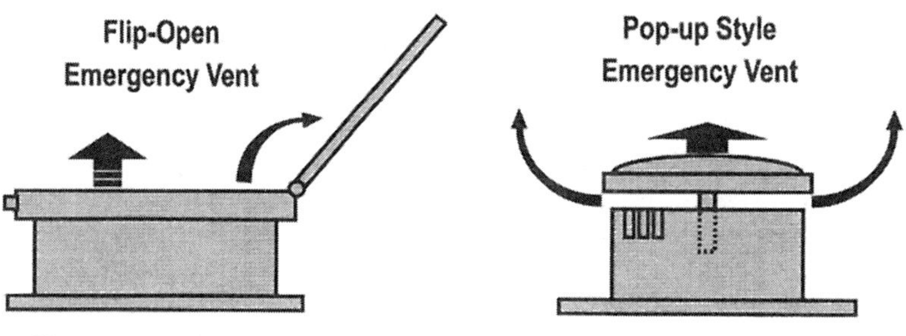

Figure 7 Emergency vents.

4. Design Considerations

There are variations in size from 3 in. to 10 in. Mounting options include male-threaded, female-threaded, and flanged. Material options include iron or aluminum bodies, metal seats, or O-ring seats, with or without screen. Certain suppliers have these vents available with a UL listing.

IV. LEGAL REQUIREMENTS

A. Special Notice

Any specification for AST systems covered under this outline must include proper accommodation for emergency venting. If a system of this type is installed and operated without proper emergency venting, a highly unsafe condition will exist. This is a condition that could cause serious injury or death.

Federal, state, and local laws provide requirements to follow in the design and specification of these AST systems. An AST specification could be subject to the mandates of a national fire code, local fire ordinance, federal environmental protection policies, local air quality standards, occupational safety considerations, local zoning laws, etc. These requirements often vary from one jurisdiction to the next. They may vary depending on other factors such as proximity to population, water supply, volume of operation, etc. When specifying equipment it is critical to have a thorough understanding of all the regulatory requirements affecting a specific AST system for a specific location.

B. Emergency Venting

Emergency venting is one aspect that is universally required by law in the United States and Canada, primarily through application of national fire codes and standards. The codes are very specific regarding proper type and size of vent for a tank. Most codes require that every vent have a flow rating permanently labeled on the vent itself so it can be verified in the field by an inspector. This rating is most commonly designated in cubic feet per hour (CFH), which normally increases with the size of the vent. An inspector may require that the vent, or the whole tank system, be listed by an independent lab such as Underwriters Laboratories. It is important to verify what the code and local fire officials expect in providing emergency venting, and that the specification meets these requirements.

The latest versions of national fire codes require three other specific equipment features on AST systems:

Overfill prevention device
Overfill alarm
A means of liquid level indication, such as a tank gauge

These new requirements are in response to spill and overfill control. An AST overfill is both an environmental and fire hazard. Product is usually pumped into ASTs. When too much product is pumped into a tank, it may reach the emergency vent and spill out. Flammable liquid spilled outside the tank creates the potential for explosion and/or fire. Equipment is available that can help prevent this from happening. The codes have recognized this and now require that this equipment be provided on these systems. Not all jurisdictions have adopted the latest versions, and in some cases these requirements may not yet be included under present law. Therefore, it is important to know the status of the law and understand whether a system can be retrofitted to meet any future requirements.

C. Overfill Prevention Valve

This device was briefly discussed before. It is a valve that is installed in the top of the tank at the fill port (Fig. 3). It can be used as a connection to the fill hose through a tight connection, or it can be installed within the piping on a remote-fill system. It is normally set to shut off the flow of product at a preset limit up to 95 percent of the tank's total liquid level capacity. Most valves have the capability of setting this limit within a range from 90 percent to 95 percent full. The valve's shutoff mechanism varies to some degree from a poppeted style to hydraulic principle. There have been cases where one type works and the other hasn't due to flow and viscosity. So it is important to research the differences and, if possible, test them before final selection.

D. Overfill Alarm

There are many types of alarms on the market. The principle is that when the liquid reaches a certain level in the tank, such as 90 percent full, an alarm (Fig. 4) will sound and hopefully alert the operator to shut off the flow into the tank. There are electronic alarms, battery operated and those that are fully mechanical. There is even a fully mechanical alarm that also serves as a normal pressure vacuum vent in combination. The probes range from float type, rod and switch, hydrocarbon sensors, and ultrasonic. The alarms are often hooked into gage systems and elaborate networks, but they are also built to stand alone.

E. Liquid Level Indication

As with underground tanks, liquid level measurement (Fig. 3) and reconciliation can be handled in a variety of ways. Product choices range from the inexpensive float-type indicator used on small tanks to elaborate electronic systems. There are a few differences between UST and AST systems. Generally, for ASTs it is not required to measure and report for leak detection purposes. Thus the level of accuracy does not need to be near that used for USTs—1/8-in. is adequate for most AST applications. Also, the fluctuation due to temperature on an AST would make a higher level of precision impractical. Secondly, ASTs vary in measurement range from a few feet on small horizontal tanks to over 30 ft on large vertical tanks. USTs generally are in the range of 12 ft. Thirdly, AST gauges employ a much greater range of mechanical devices, which causes installation to become a much greater variable factor in the total cost of the gauge.

1. Options

The simplest gauges are those that mount on the top of a small tank and have a 1/4, 1/2, 3/4, and Full indicator. These gauges are popular because of their cost and may be adequate for the purpose intended. The next step is to start indicating liquid level via feet and inches on a mechanical tape or dial. These gauges rely on a float attached to a wire. They operate simply: as the float moves up or down, the indicator moves correspondingly. These gauges vary in cost, operation, quality, performance, and ease of installation. This area of equipment probably requires as much research as any to become well versed on the options, benefits, and new developments. Electronic gauging for AST systems reflects a growing trend with less expensive, highly reliable means of automating basic functions. Traditionally, an electronic setup was available and has been used in process engineering for many years. These systems are expensive and for the most part customized to the process. There is a growing trend for more universal measurement options and total communication via modems, fax, etc., for the future management of product inventory.

F. Antisiphon

Antisiphon is another provision required by codes on AST systems. This feature (Fig. 3) provides the means for preventing liquid from siphoning out of a tank when a line connected to the tank has been damaged or fractured downstream below the liquid level. Antisiphon options were covered earlier in the chapter.

V. SUMMARY

The specifier must understand the full picture of requirements and desired objectives before attempting to specify any system or components. There are certain basics to cover, but the true beauty of a successful system will only be realized when the specifier goes beyond the basics in researching equipment for the options and benefits best suited to the needs of an individual site. There are certain life safety considerations such as emergency venting that are fundamental, and any specification will be held accountable for proper resolution of these issues.

The future is ever changing and full of new opportunities for equipment to affect the way we handle operations, especially in the area of communications and inventory management. Specifiers need to be up-to-date on these changes to maximize value and safety in the service life on any AST system.

Appendix

I. UNDERGROUND STORAGE TANKS

A. Regulations Potentially Applicable to USTs

1. U.S. EPA Regulations

Resource Conservation and Recovery Act (RCRA) Subtitle I—regulations addressing USTs storing petroleum and hazardous substances (40 CFR 280)
RCRA Subtitle C—hazardous waste regulations (40 CFR 261 through 265)
Stage II Vapor Recovery Regulations—established in the Clean Air Act Amendments of 1990 (Section 182(b)(3)(A)
Spill Prevention, Control, and Countermeasures (SPCC) within the Clean Water Act (40 CFR 112)
Clean Air Act Title V—operating permits
Emergency Planning & Community Right-to-Know Act (EPCRA) (Title III of Superfund Amendments and Reauthorization Act)
Used-Oil Management Standards (40 CFR 279)

2. OSHA Regulations

Hazardous Waste Operations and Emergency Response Regulations (Hazwoper) (29 CFR 1910.120)
Flammable and Combustible Liquids Standard (29 CFR 1910.106)
Hazard Communication Standard (HCS) (29 CFR 1910.1200)
Confined Space Safety Standard (29 CFR 1910.146)

B. Codes Applicable to USTs

National Fire Protection Association
NFPA30 (Flammable and Combustible Liquids Code)

NFPA 30A (Automotive and Marine Service Station Code)

International Conference of Building Officials
UFC Article 52 (Motor Vehicle Fuel-Dispensing Stations)*
UFC Article 79 (Flammable and Combustible Liquids)*

Building Officials and Code Administrators International
BOCA National Fire Prevention Code*

Southern Building Code Congress International
SBCCI Standard Fire Prevention Code*

C. Standards Applicable to Steel USTs

Underwriters Laboratories, Inc.
UL 58 (Steel Underground Tanks for Flammable and Combustible Liquids)
UL 1746 (External Corrosion Protection Systems for Underground Storage Tanks)
Part I—Preengineered Cathodic Protection Systems
Part II—Composite Tanks
Part III—Jacketed Tanks
Part IV—Coated Tanks1**

Underwriters' Laboratories of Canada
CAN/ULC S603 (Underground Steel Tanks)
CAN/ULC S603.1 (Galvanic Corrosion Protection Systems for Underground Steel Tanks)
ULC S616 (Liquid Protective Coating Materials as Required by ULC S603.1 for Use in Connection with the Corrosion Protection of Underground Tanks)
ULC ORD-C58.20 (Special Corrosion Protection Underground Tanks)
ULC ORD-C58.10 (Underground Jacketed Steel Tanks)
CAN/ULC S618 (Magnesium and Zinc Anode Assemblies and Zinc Reference Electrodes)

---

*As of November 1998, UFC, BOCA and SBCCI codes are slated to be unified into the new International Fire Code (IFC), scheduled for first edition publishing in 2000. The umbrella organization for the IFC is the International Code Council (ICC). See the contact appendix for ICC information. However, the Western Fire Chiefs Association (WFCA) has obtained rights to continue the Uniform Fire Code, independent of the ICBO. Therefore, the UFC will likely remain a code followed in certain regions of the country. While still under negotiation at this writing, contact information for the WFCA also is included in the contact appendix.
**This is a pending designation, as of November 1998.

CAN4-S631 (Isolating Bushings for Steel Underground Tanks Protected with Coatings and Galvanic Systems)

Steel Tank Institute
STI F841 (Standard for Dual-Wall Underground Steel Storage Tanks)
STI F894 (ACT-100 Specification for External Corrosion Protection of FRP Composite Steel Underground Storage Tanks)
STI F922 (Permatank Specification)
STI-P3 (STI-P3 Specification and Manual for External Corrosion Protection of Underground Steel Storage Tanks)
STI F961 (ACT-100U Specification for External Corrosion Protection of Composite Steel Underground Storage Tanks)

D. Recommended Practices Applicable to Steel USTs

American Petroleum Institute
API RP-1615 (Installation of Underground Petroleum Product Storage Systems)
API RP-1632 (Cathodic Protection of Underground Petroleum Storage Tanks and Piping)

NACE International
NACE RP-0169 (Control of External Corrosion on Underground or Submerged Metallic Piping Systems)
NACE RP-0285 (Corrosion Control of Underground Storage Tank Systems by Cathodic Protection)

American Society for Testing and Materials
ASTM E1430-91 (Standard Guide for Using Release Detection Devices with Underground Storage Tanks)
ASTM E1526-93 (Standard Practice for Evaluating the Performance of Release Detection Systems for Underground Storage Tank Systems)

Petroleum Equipment Institute
PEI RP-100 (Recommended Practices for Installation of Underground Liquid Storage Systems)

Steel Tank Institute
STI R821 (STI-P3 Installation Instructions)
STI R831 (Optional Recommended Practice for Control of Localized Corrosion Within Underground Steel Petroleum Storage Tanks)
STI R891 (Recommended Practice for Hold Down Strap Isolation)
STI R913 (ACT-100 Installation Instructions)
STI R923 (Permatank Installation Instructions)
STI R971 (ACT-100U Installation Instructions)

II. ABOVEGROUND STORAGE TANKS

A. Regulations Potentially Applicable to ASTs

1. EPA Regulations

Spill Prevention, Control and Countermeasures (SPCC) within the Clean Water Act (40 CFR 112)
Oil Pollution Act (OPA) of 1990
National Pollutant Discharge Elimination System (NPDES)
Clean Air Act of 1970, plus amendments of 1977 and 1990
Superfund Amendments and Reauthorization Act (SARA) Title III
Comprehensive Environmental Response, Compensation and Liability Act (CERCLA) (Superfund)
Resource Conservation and Recovery Act (RCRA) Subtitle C—hazardous waste regulations (40 CFR 261 through 265)

2. OSHA Regulations

Hazardous Waste Operations and Emergency Response Regulations (Hazwoper) (29 CFR 1910.120)
Flammable and Combustible Liquids Standard (29 CFR 1910.106)
Hazard Communication Standard (HCS) (29 CFR 1910.1200)
Confined Space Safety Standard (29 CFR 1910.146)
Benzene (29 CFR 1910.1028)
Process Safety Management of Highly Hazardous Chemicals; Explosives and Blasting Agents (PSM standard) (29 CFR 1910.119)

B. Codes Applicable to ASTs

National Fire Protection Association
NFPA30 (Flammable and Combustible Liquids Code)
NFPA 30A (Automotive and Marine Service Station Code)

International Conference of Building Officials
UFC Article 52 (Motor Vehicle Fuel-Dispensing Stations)*
UFC Article 79 (Flammable and Combustible Liquids)*
UFC Appendix II-F (Protected Aboveground Storage Tanks for Motor Vehicle Fuel-Dispensing Stations Outside Buildings)*
UFC Appendix II-K (Nonprotected Aboveground Steel Tanks for Private Motor Vehicle Fuel-Dispensing Stations Outside Buildings)

Building Officials and Code Administrators International
BOCA National Fire Prevention Code*

Southern Building Code Congress International
SBCCI Standard Fire Prevention Code*

C. Standards Applicable to Steel ASTs

Underwriters Laboratories, Inc.
UL 142 (Steel Aboveground Tanks for Flammable and Combustible Liquids)
UL 2085 (Protected Aboveground Tanks for Flammable and Combustible Liquids)
UL Subject 2080 (Outline of Investigation for Fire Resistant Aboveground Tanks for Flammable and Combustible Liquids)
UL Subject 2244 (Standard for Aboveground Flammable Liquids Tank Systems)

Underwriters' Laboratories of Canada
ULC S601 (Aboveground Horizontal Shop Fabricated Steel Tanks)
ULC S630 (Aboveground Vertical Shop Fabricated Steel Tanks)
ULC S652 (Tank Assemblies for Collection of Used Oil)
ULC S653 (Contained Aboveground Steel Tank Assemblies)
CAN/ULC S602 (Aboveground Steel Tanks for Fuel Oil and Lubricating Oil)
CAN/ULC S643 (Aboveground Shop Fabricated Steel Utility Tanks)
ULC ORD-C142.5 (Aboveground Concrete Encased Steel Tank Assemblies)
ULC ORD-C142.16 (Aboveground Protected Tank Assemblies)
ULC ORD-C142.18 (Aboveground Rectangular Steel Tanks)
ULC ORD-C142.20 (Aboveground Secondary Containment Tanks)
ULC ORD-C142.23 (Aboveground Waste Oil Tanks)
ULC ORD-C142.21 (Aboveground Used Oil Systems)
ULC ORD-C142.22 (Contained Aboveground Vertical Steel Tank Assemblies)

Steel Tank Institute
STI F911 (Standard for Diked Aboveground Storage Tanks)
STI F921 (Standard for Aboveground Tanks with Integral Secondary Containment)
STI F941 (Fireguard Thermally Insulated Aboveground Storage Tank Standard)

D. Recommended Practices Applicable to Steel ASTs

Petroleum Equipment Institute
PEI RP200 (Recommended Practices for Installation of Aboveground Storage Systems for Motor Vehicle Fueling)

Steel Tank Institute
STI R893 (Recommended Practice for External Corrosion Protection of Shop-Fabricated Aboveground Storage Tank Floors)
STI R931, F921 (Installation Instructions)
STI R912 (Installation Instructions for Factory Fabricated Aboveground Tanks)

STI R942 (Fireguard Installation & Testing Instructions for Thermally Insulated, Lightweight, Double Wall Fireguard Aboveground Storage Tanks)

III. RELATED INFORMATION

A. Codes

National Fire Protection Association
NFPA 31 (Installation of Oil-Burning Equipment)
NFPA 329 (Recommended Practice for Handling Underground Releases of Flammable and Combustible Liquids)
NFPA 395 (Storage of Flammable and Combustible Liquids on Farms and Isolated Construction Projects)

International Conference of Building Officials
UFC Appendix II-B (Protection of Flammable and Combustible Liquids in Locations Subject to Flooding)
UFC Appendix II-J (Storage of Flammable and Combustible Liquids in Tanks Located Within Below-Grade Vaults)

B. Standards

Underwriters Laboratories, Inc.
UL 1316 (Glass-Fiber-Reinforced Plastic Underground Storage Tanks for Petroleum Products, Alcohol and Alcohol-Blended Mixtures)
UL 971 (Nonmetallic Underground Piping for Flammable Liquids)
UL 80 (Steel Inside Tanks for Oil-Burner Fuel)
UL 443 (Steel Auxiliary Tanks for Oil-Burner Fuel)
UL Subject 2245 (Outline of Investigation for Below-Grade Vaults for Flammable Liquid Storage Tanks)
UL 525 (Flame Arresters for Use on Vents of Storage Tanks for Petroleum Oil and Gasoline)

Underwriters' Laboratories of Canada
ULC S615 (Underground Reinforced Plastic Tanks)
ULC S633 (Flexible Underground Hose Connectors for Flammable and Combustible Liquids)
ULC ORD-C107.4 (Ducted Flexible Underground Piping Systems for Flammable and Combustible Liquids)
ULC ORD-C107.7 (Glass-Fibre-Reinforced Plastic Pipe and Fittings)
ULC ORD-C107.12 (Line Leak Detection Devices for Flammable Liquid Piping)
ULC ORD-C107.19 (Secondary Containment of Underground Piping)
ULC ORD-C107.21 (Under-Dispenser Sumps)

ULC ORD-C58.12 (Leak Detection Devices [Volumetric Type] for Underground Storage Tanks)
ULC ORD-C58.14 (Leak Detection Devices [Non Volumetric Type] for Underground Storage Tanks)
ULC ORD-C58.15 (Overfill Protection Devices for Underground Tanks)
ULC ORD-C58.19 (Spill Containment Devices for Underground Tanks)
ULC ORD-C58.9 (Secondary Containment Liners for Underground and Aboveground Tanks)
ULC ORD-C142.19 (Spill Containment Devices for Aboveground Tanks)
ULC ORD-C525 (Flame Arresters for Use on Vents of Storage Tanks for Petroleum Oil and Gasoline)

American Society of Mechanical Engineers
ANSI/ASME B31.3 (Process Piping)
ANSI/ASME B31.4 (Liquid Transportation Systems for Hydrocarbons, Liquid Petroleum Gas, Anhydrous Ammonia and Alcohols)

American Petroleum Institute
API Spec 12B (Specification for Bolted Tanks for Storage of Production Liquids)
API Spec 12D (Specification for Field-Welded Tanks for Storage of Production Liquids)
API Spec 12F (Specification for Shop-Welded Tanks for Storage of Production Liquids)
API Spec 12P (Specification for Fiberglass-Reinforced Tanks)
API Standard 620 (Design and Construction of Large, Welded, Low-Pressure Storage Tanks)
API Standard 650 (Welded Steel Tanks for Oil Storage)
API Standard 2000 (Venting Atmospheric and Low-Pressure Storage Tanks [Nonrefrigerated and Refrigerated])
API Standard 2610 (Design, Construction, Operation, Maintenance and Inspection of Terminal and Tank Facilities)

American Society for Testing and Materials
ASTM D 1998 (Polyethylene Upright Storage Tanks)
ASTM D 3299 (Filament-Wound Glass-Fiber-Reinforced Thermoset Resin Chemical-Resistant Tanks)
ASTM D 4097 (Contact-Molded Glass-Fiber-Reinforced Thermoset Resin Chemical-Resistant Tanks)

C. Recommended Practices

Petroleum Equipment Institute
PEI RP300 (Installation/Testing of Vapor Recovery Systems at Vehicle Fueling Sites)

Steel Tank Institute
STI R892 (Recommended Practice for Corrosion Protection of Underground Piping Networks Associated with Liquid Storage and Dispensing Systems)

American Petroleum Institute
API RP2350 (Overfill Protection for Petroleum Storage Tanks)

IV. CONTACTS

American Iron and Steel Institute
1101 17th St, NW, Suite 1300
Washington D.C., 20036-4700
Telephone: 202-452-7100
Fax: 202-463-6573
Website: www.steel.org

American National Standards Institute
11 West 42nd St.
New York, NY 10036
Telephone: 212-642-4900
Fax: 212-398-0023
Website: web.ansi.org

American Petroleum Institute
1220 L Street, Northwest
Washington, DC 20005
Telephone: 202-682-8000
Fix: 202-962-4776
Website: www.api.org

American Society for Testing and Materials
100 Barr Harbor Drive
West Conshohocken, PA 19428-2959
Telephone: 610-832-9585
Fax: 610-832-9555
Website: www.astm.org

Building Officials and Code Administrators International
4051 W. Flossmoor Rd.
Country Club Hills, IL 60478
Telephone: 708-799-2300
Fax: 708-799-4981
Website: www.bocai.org

California Air Resources Board
2020 L Street
Sacramento, CA 95814
Telephone: 916-322-2990
Fax: 916-445-5025
Website: www.arb.ca.gov

U.S. Environmental Protection Agency
Website: www.epa.gov/swerust1

U.S. EPA, Region 1
1903 John F. Kennedy Building
Mailcode: HBO
Boston, MA 02203-2211
Phone: 617-573-9604
Fax: 617-573-9662

U.S. EPA, Region 2
Water Compliance Branch
290 Broadway
Code: 2DECA-WCB
New York, NY 10007-1866
Phone: 212-637-4232
Fax: 212-637-3901

U.S. EPA, Region 3
1650 Arch Street
Mailcode: 3WC21
Philadelphia, PA 19103-2029
Phone: 215-814-3231
Fax: 215-814-3163

U.S. EPA, Region 4
Water Management Division
Underground Storage Tank Section
Atlanta Federal Center
61 Forsyth Street, SW
Atlanta, GA 30303-3104
Phone: 404-562-9424
Fax: 404-562-9439

U.S. EPA, Region 5
77 W. Jackson Boulevard
Mailcode: DRU-7J
Chicago, IL 60604-3590

Phone: 312-886-6136
Fax: 312-353-3159

U.S. EPA, Region 6
First Interstate Bank Tower
1445 Ross Avenue, Suite 1200
Mail Code: 6PD-U
Dallas, TX 75202-2733
Phone: 214-665-6760
Fax: 214-665-7263

U.S. EPA, Region 7
ART/GUTS
726 Minnesota Avenue
Kansas City, KS 66101
Phone: 913-551-7651
Fax: 913-551-7947

U.S. EPA, Region 8
999 18th Street, Suite 500
Mail Code: 8P2-W-GW
Denver, CO 80202-2466
Phone: 303-312-6137
Fax: 303-312-6741

U.S. EPA, Region 9
75 Hawthorne Street
Mail Code: WST-8
San Francisco, CA 94105
Phone: 415-744-2083
Fax: 415-744-1044

U.S. EPA, Region 10
1200 Sixth Avenue
Mail Code: WD-133
Seattle, WA 98101
Phone: 206-553-2857
Fax: 206-553-1280

International Code Council
5203 Leesburg Pike, Suite 708
Falls Church, VA 22041
Telephone: 703-931-4533
Fax: 703-379-1546
Website: www.intlcode.org

International Conference of Building Officials
5360 Workman Mill Rd.
Whittier, CA 90601-2298
Telephone: 562-699-0541
Fax: 562-692-6031
Website: www.icbo.org

NACE International
1440 South Creek Dr.
Houston, TX 77084-4906
Telephone: 281-228-6200
Fax: 281-228-6300
Website: www.nace.org

National Fire Protection Association
1 Batterymarch Park
Quincy, MA 02269
Telephone: 617-770-3000
Fax: 617-770-0700
Website: www.nfpa.org

Petroleum Equipment Institute
P.O. Box 2380
Tulsa, OK 74101-2380
Telephone: 918-494-9696
Fax: 918-491-9895
Website: www.peinet.org

Southern Building Code Congress International
900 Montclair Rd.
Birmingham, AL 35213-1206
Telephone: 205-591-1853
Fax: 205-591-0775
Website: www.sbcci.org

Steel Tank Institute
570 Oakwood Rd.
Lake Zurich, IL 60047
Telephone: 847-438-8265
Fax: 847-438-8766
Website: www.steeltank.com

Underwriters Laboratories, Inc.
333 Pfingsten Rd.
Northbrook, IL 60062-2096

Telephone: 847-272-8800
Fax: 847-272-8129
Website: www.ul.com

Underwriters' Laboratories of Canada
7 Crouse Rd.
Scarborough, Ontario M1R 3A9 Canada
Telephone: 416-757-3611
Fax: 416-757-1781
Website: www.ulc.ca

U.S. Department of Labor
Occupational Safety and Health Administration (OSHA)
Directorate of Construction
200 Constitution Ave., NW
Washington, DC 20210
Telephone: 202-693-2345
Website: www.osha.gov

Western Fire Chiefs Association
300 N. Main St., Suite 25
Fallbrook CA 92028
Telephone: 760-723-6911
Fax: 760-723-6912
Website: www.wfca.com

Index

Aboveground storage system (AST), 21, 70–89, 91, 93, 98, 114, 185, 189, 196, 199, 219, 293, 300, 306, 308–309, 312, 331
components of, 287–288, 300–308
anti-siphon device, 15, 77, 80, 93, 196, 207, 270–271, 300, 302, 304, 306, 315, 317, 324
atmospheric vent, 76, 302
bollards, 64, 80, 93, 282–283
dispenser, 80–81, 82, 85, 93, 96–97, 197, 201, 270, 276–282, 286–287, 304, 306, 312, 315, 317–318
emergency vent, 15, 17, 45, 76, 93, 96–97, 186, 189, 194, 196, 203, 212, 226, 233, 237, 254, 256, 258–267, 284, 286–288, 302, 309, 319, 321–323, 325
fill port, 312–315, 323
flame arrestor, 76, 257, 263
[Aboveground storage system]
ladder, 196, 302
leak detection, 13, 78, 88, 97, 189, 201, 212, 245, 287, 304, 306, 308, 323
liquid-level indication, 77–78, 81, 196, 201, 322–324
normal vent, 96, 189, 194, 196, 256–257, 265–266, 287, 309, 319–320
overfill prevention, 15, 78, 191, 196, 206–207, 225, 267–269, 287, 312, 315, 322, 323
pressure-vacuum vent, 76, 257–258, 319–320
remote pump, 287, 304, 306
spill container, 77, 97, 201, 203, 206–207, 312, 315, 333
submersible turbine pump, 81, 277
suction pump, 277–278, 300, 302, 306

[Aboveground storage system]
vapor recovery, 74, 76–77, 79–81, 223, 327
design, 183–191
cylindrical, 74, 184, 194, 206, 209, 240, 242, 249
diked, 15–16, 38, 83, 97, 183, 185–186, 194–195, 207, 209–211, 214, 223, 230–232, 236, 249, 253–254, 268, 298–299, 331
double wall, 16–17, 42, 44, 84, 97, 183–184, 186, 202–203, 206–207, 211–214, 299, 302, 309, 332
fire-resistant, 14, 16–17, 75, 97, 184, 187–188, 193, 195–196, 233–234, 254–256, 277, 283, 295, 331
horizontal, 45, 97, 184, 194, 200, 202, 206–207, 212, 214, 240, 244, 248–250, 259–262, 298–299, 309, 324, 331
multi-hazard, 184, 188, 233
obround, 3
protected, 14, 16–17, 75, 96, 184, 187–189, 193, 195–197, 200, 208, 233–234, 254–256, 261, 277, 283, 330, 331
rectangular, 16, 33, 97, 184, 194, 200, 206–207, 240–241, 243, 247–248, 261–262, 331
secondary containment, 14–15, 42, 47, 75–76, 78, 93, 97, 115, 185–187, 189, 194–196, 199–200, 202–203, 208–212, 215, 219, 222–223, 225–226, 230, 233, 240–243, 245–255, 263–264, 268, 284, 286, 331
single wall, 41–42, 183–186, 194, 202–203, 206, 210, 232, 245, 253, 264, 299, 309

[Aboveground storage system]
tertiary containment, 187
vertical, 45, 96–97, 184, 194, 200, 202–203, 207, 212, 213, 240–241, 243–244, 251, 261–262, 264, 298–299, 309, 317, 324, 331
inspection, 93, 226, 237, 266, 284–288, 299, 310
installation, 72, 83–89, 98, 189, 196, 214, 225, 227, 234, 238, 244, 282–286, 288, 293–310, 321, 324, 331–332
backfill, 84–86, 87
foundation, 83, 86, 225, 244–247, 279, 283, 295–298
grounding, 276, 288, 300
permit, 93, 294
separation distance, 52, 96, 228–232, 234, 255, 295
setback requirements, 191
tank, 2, 13–17, 33, 41–42, 45–47, 51–53, 70, 72–81, 83–89, 91, 92, 93, 95–98, 114, 118, 183–189, 191, 193–197, 199–203, 206–215, 217–223, 225–228, 230–234, 236–237, 239–240, 242–249, 251–252, 255–268, 270, 274–278, 281–288, 293–296, 298–300, 302, 304, 306, 308–312, 315, 317–325, 330–332
bulkheads, 43, 184
compartment, 43, 95, 202, 214, 260
fire rated, 70, 75, 97, 187, 214
for standby generator, 197, 233–234, 285, 306
hidden costs, 93
insulated, 16–17, 72–73, 75, 84, 97, 183, 195–196, 208, 254, 295, 332

[Aboveground storage system]
nameplates, 226, 266, 284–285
primary, 42, 44–45, 78, 88, 184, 186, 188–189, 194–197, 203, 207–208, 210, 212–213, 239, 253–254, 263–264, 284, 299, 309
rain shield, 183, 186, 211
testing, 88, 189, 233, 309–310, 332
documentation, 87, 89
hydrostatic, 89, 194–195, 206–207, 283, 286–287
observation of, 87, 288, 310
pneumatic, 286
radiographic, 287
system components, 74, 87–89
Abrasive blasting, 38
ACT-100, 6, 12, 139, 162, 328, 329
ACT-100-U, 141, 162, 329
Air test
at installation site, 146–147, 150–151, 286, 309
at plant, 37–38
safety relief device, 38, 147
American Iron & Steel Institute, 100, 334
American National Standards Institute, 51, 105, 110–111, 238, 334
American Petroleum Institute (API), 3, 11, 51, 55, 72, 123, 144, 218, 220, 238, 256, 266, 268, 273, 329, 333–334
report on AST spills, 267
American Society for Testing and Materials (ASTU), 34, 51, 58, 65, 75, 84, 105, 145, 238, 244, 270, 273, 329, 333–335
American Society of Mechanical Engineers, 101, 269, 272–274, 286, 287, 333
American Welding Society, 274
Annulus (*see* Interstitial space)
Anode, 7, 45–46, 58, 95, 109, 115, 119, 138, 157, 159, 161–162, 166–168, 171–172, 299, 328
Authority having jurisdiction, 12, 17, 44, 55, 70, 88–89, 92, 96, 107, 147, 185, 189, 191, 202, 214, 225, 272, 309

Ballast, 148
Battelle Laboratories, 100–101, 104
Bench tank, 184
Buckling, 46, 100–103, 108
Building code, 21–22, 24, 26, 30–31, 73, 83, 183, 185, 189, 217, 244, 310
Building Officials and Code Administrators International, Inc. (BOCA), 17, 25–27, 73, 77–78, 80–81, 83, 226, 268, 328, 330, 335

California Air Resources Board, 81, 223, 335
California Fire Chiefs Association, 31
Canadian Petroleum Products Institute, 51
Cathodic protection, 7–9, 12, 46, 51, 58–59, 95, 106–107, 109–110, 117, 123–124, 137, 138, 141, 144, 146, 150, 152, 154, 156–157, 161–174, 176–177, 271, 300, 328–329
AST bottom, 46, 299–300
continuity test, 150, 166, 169–170
impressed current, 150, 170, 299
monitoring , 95, 157, 159, 162, 170
test station, 58–59, 60, 165–166, 168
Chemical storage , 17, 42, 52, 95
Cleanup (*see* Remediation)
Coating, 33–34, 39–40, 45, 47, 58,

[Coating]
74, 106–109, 115, 119, 138, 147–149, 151, 154, 166, 172, 176–177, 244, 270–271, 328
AST paints, 40, 46
coal-tar epoxy, 7, 39–40, 138
dielectric, 7, 46, 58, 119, 176
fiberglass-reinforced plastic, 6, 8, 39–40, 46, 59, 99–102, 107, 138–139, 144, 147, 154
polyurethane, 7, 39–40, 137–138, 141, 146–147, 150
Composite tanks, 12, 39–40, 95, 100–101, 109–110, 137–139, 144, 146, 160, 171, 328
Comprehensive Environmental Response Compensation and Liability Act, 95, 330
Concrete pad, 145, 296
Copper Sulfate Electrode ([CSE] *see* Reference electrode)
Corrosion, 3–6, 9–11, 84, 100, 106–107, 109, 117–119, 121, 123, 138–140, 153, 155, 157, 161–174, 176, 271, 272, 299, 308–309, 329
Corrosion protection, 8, 11, 39, 41, 46, 52, 84, 95, 100, 106–107, 109, 114–115, 117–119, 121, 124, 138, 140–141, 153, 157, 168, 176, 271–272, 299–300, 308, 328, 331, 334

DOT (*see* U.S. Department of Transportation)
Dow Chemical, 7
Dye penetrant test, 38, 195

Electrical isolation, 7, 138–139, 146, 150–151, 154, 169
Electrolyte, 161–162
Environmental Defense Fund, 218
Environmental Protection Agency ([EPA] *see* U.S. Environmental Protection Agency)

F911, 210–211, 214, 220
F921, 16–17, 212, 214, 220
Fabrication, 41, 45, 52, 55
Fiberglass-reinforced plastic (FRP), 5–6, 12, 42, 46, 54, 71, 82, 95, 100, 105, 107, 114, 119, 121, 123, 171, 181, 332–334
Field erected, 2, 15, 46, 232
Fire code, 8, 11, 13–15, 21, 22, 29, 30–31, 43, 45, 51, 64, 80, 83–84, 93, 96, 98, 183, 185, 186–189, 214, 217, 226–227, 232–235, 237–238, 245–246, 248, 253, 255–256, 260, 266–268, 279, 281–283, 286–287, 300, 320, 322
Firefighting
fire-department access, 225, 237
fire flow, 238–240, 242
fire width calculation, 231
foam, 237, 239–240, 242–244, 260
Fireguard, 17, 73–74, 84, 89, 331–332
Flammable vapor, 226, 234, 236, 239, 267, 274–276
Flanged head, 35

Gadsden, AL, 233
Galvanic series, 162, 172
Gasoline, 43, 53, 74, 116–117, 155, 178, 201–203, 229, 240, 243–244, 260, 275, 277, 281, 286, 288, 319, 332–333
General Accounting Office, 219
Generator-base tank, 197, 206–207, 233
Groundwater, 5–6, 46, 63, 82, 99, 107, 153, 178, 219, 256

Half cell (*see* Reference electrode)
Heating oil, 3–4, 122, 185
Hose-stream resistance, 75
Holiday, 119, 147, 151
  test, 39, 40, 109, 139, 144, 147, 151
Hydrocarbon, 2, 3, 238, 261, 287–288, 323, 333
Hydrostatic load test, 99

Inspection, 33, 40
Installation (*see* Aboveground storage system and Underground storage system)
Institute of Makers of Explosives, 283
Insurance Rating Bureau 30
Interference, 173
Internal corrosion, 3, 59, 117, 138, 155
Internal lining, 8, 12, 47, 52, 171
International Code Council, 27, 337
International Conference of Building Officials, 25–27, 31–32, 327, 330, 332, 337
  International Building Code, 26
  Uniform Building Code, 26, 31, 75
International Electrotechnical Commission, 105
International Fire Code Institute, 25–26, 32, 55, 73
  International Fire Code, 226
Interstitial space, 9, 42–45, 54, 58, 62, 71, 75, 107, 110, 119, 146, 176, 178, 203, 253–254, 263, 286, 309

Jacketed tank, 12, 38, 39–40, 42, 45, 47, 107, 109–110, 115, 117–119, 137, 140, 144, 146, 328
Jet fire, 239
Joggle joint (*see* Offset joint)

Kansas City, KS, 259
Kennedale, TX, 259
Kerosene, 43, 229
Knuckle radius, 35

Leak detection, 10–11, 41–43, 46, 50, 52, 62, 107, 115, 123–124, 138–139, 177–179, 201
  piping options, 62, 124, 179
  tank options, 43, 62, 124, 178–179
Lifting lug strength test, 108, 206

Magnesium, 5, 7, 58, 115, 119, 161–162, 171, 328
Manhole, 62, 68–69, 82, 86, 89, 118–119, 179–180, 203, 321
Manway, 35, 39–40, 59, 76, 184, 265
Modulus of elasticity, 100, 108
Motor oil, 178, 197

NACE International, 9, 52, 56, 72, 162, 271, 299–300, 329, 337
National Board of Fire Underwriters, 2, 29, 108, 194
National Electrical Code (NEC) (NFPA 70), 55, 60, 67, 73, 81, 86, 274–276, 278, 287, 288
National Fire Code (U.S.), 29–31
National Fire Code of Canada, 202, 207
National Fire Protection Association (NFPA), 2, 5, 8–9, 14–17, 25, 29–30, 51, 55, 60, 73, 77, 80–81, 83, 84, 89, 93, 96, 103, 105–108, 111, 121, 143, 151, 187–188, 196, 202, 214–215, 226, 232–233, 238, 240, 242, 249, 253, 255–256, 260, 263, 266–268, 270, 272–275, 278–279, 281, 284, 286, 308, 327, 330, 332, 337
National Leak Prevention Association, 12

National Steel Tank Association, 1
National Weather Service rainfall data, 248

OSHA (*see* U.S. Department of Labor, Occupational Safety and Health Administration)
Offset joint, 35
Oil-water separator, 139
Overfill containment, 52

Permatank, 12, 140, 328–329
Petroleum, 1, 3, 5, 10, 17, 33, 43, 50–51, 53, 55–57, 70, 72, 74, 95, 105, 121–122, 140, 144, 154, 167, 177, 183, 193, 202, 209–210, 221, 237, 256, 268, 269, 277–278, 329, 332–333
Petroleum Equipment Institute (PEI), 3, 5, 11–12, 50–51, 55, 69, 72, 84, 88, 98, 123, 143, 146–148, 151, 220, 293–294, 329, 331, 334, 337
Petroleum Marketers Association of America, 218
Piping, 59, 61–63, 65–66, 68, 70, 77–80, 82, 84–85, 87–89, 93, 96–97, 110, 116, 119, 139, 144, 146, 150, 154–155, 166, 169–170, 173, 175, 177, 197, 219–220, 225–226, 238, 267–274, 278, 283, 286–287, 295, 302, 306, 308–309, 312, 315, 317–318, 320, 323, 329, 333
  aboveground, 78, 196, 269, 274, 308–309, 318
  block valve, 80, 93, 302, 304, 306, 309, 315
  check valve, 93, 175–176, 179, 302, 315, 317
  containment sump, 62–63, 81–82, 85, 87, 89, 116, 176–177, 181, 308
  double wall,175–177
  emergency shutoff valve, 81, 279–280, 287–288, 315, 317
  fiberglass-reinforced plastic, 12, 61, 64, 79–80, 116, 177, 201, 274, 333
  fire valve, 302, 306
  flexible, 12, 63–64, 79–80, 116, 176–177, 201, 308, 333
  flow restrictor, 179
  leak detection, 78, 98, 116, 123–124, 201, 304, 306, 308, 333
  line strainer, 315, 318
  monitoring, 47, 176
  pressure-regulating valve, 302–303
  pressure-relief valve, 80, 278, 304, 306, 315, 318–319
  pressurized, 11, 175–176, 179, 304
  return line, 176, 179, 308
  secondary containment, 10, 13, 61, 79, 88, 116, 181, 201, 333
  single wall, 175–176
  suction, 124, 175–177, 179
  steel, 12, 61, 64, 271, 273, 299, 308, 310
  tightness test, 61, 179, 310
  underground, 56, 73, 79, 176, 177, 256, 271, 299, 302, 304, 308, 332, 334
  vapor recovery, 57, 61, 79–80, 176–177, 179–180, 327
  vent, 10, 61, 63, 68, 79, 87–88, 97, 176–177, 179
Product compatibility, 52, 58, 84, 119, 123, 138, 140, 177, 278
Public Law, 122, 153, 154

Quality assurance, 7, 138
Quality control, 7, 33–34, 138–141, 160

Reese, Dr. R. Allan, 100–101, 104
Reference cell (*see* Reference electrode)
Reference electrode, 59, 115, 119, 157, 163–168, 170–173, 328
  buried, 166
  remote, 167
Release detection (*see* Leak detection)
Remediation, 47
Resource Conservation and Recovery Act, 122, 154
Roark formula, 102–103, 108

60 Minutes, 9
Seismic requirements, 75, 83, 86
Service station, 1–2, 4–5, 13–15, 17, 25, 42, 55, 73, 191, 255, 259, 276, 281, 288, 327, 330
Southeast Association of Fire Chiefs, 25–26
Southwest Association of Fire Chiefs, 25, 26
Southern Building Code Congress International, Inc. (SBCCI), 17, 25–27, 73, 268, 328, 331, 337
Southwest Research Institute (SwRI), 96, 184, 188, 233
Specification, 49, 53, 71–72, 311
  performance-based, 49–50, 53
  prescriptive, 49–50, 53
  samples, 53–89
Spill prevention, 52, 92, 124
Standard, defined, 21
Steel, 99, 105–106, 109, 114–115, 117, 121, 140–141, 144, 147, 151, 153, 161–162, 171–172, 181, 184, 186, 194, 197, 200, 202, 206–207, 239, 254, 258–259, 270, 284, 299, 308, 310, 317, 328–329, 331
  carbon equivalency formula, 34
  stainless, 6, 45, 60, 63, 80, 82
  surface preparation, 38
  wall thickness, 99–103
Steel tank baggie, 5
Steel Tank Institute (STI), 1, 4, 6–8, 11–13, 35, 50, 52–54, 56, 59, 71, 73, 75–76, 84, 88–89, 99–100, 102–104, 137–139, 141, 157, 210, 212, 214, 293–294, 299, 328–329, 331, 334, 338
  Midwest 56 standard, 4
  QuickSpec software, 52, 89
  STI-86, 13
  STI-LIFE, 6, 138
  sti-P3, 7, 11, 53, 56–57, 60, 89, 95, 138, 144, 154, 157, 159, 162, 328
Striker plate, 138
Structural failure, 10, 155
Structural integrity, 5, 11, 46, 100–101, 103, 160, 187, 239
Suffolk County (NY), 11, 155

Tank Talk, 103–104
Tillinghast, 157, 159
Two-hour fire, 14, 16, 75, 97, 255

Underground storage system, Tank (UST), 21, 51, 53–70, 91, 95–96, 98, 114, 121–123, 125, 153–155, 183, 193, 329
  closure, 124–125
  containment sump, 13, 59, 63, 65–66, 68–70, 181
  dispenser, 1, 49, 52, 60, 62–64, 66
  emergency shutoff, 60–61
  excavation liners, 10
  installation, 11, 56, 60, 65–70, 95, 116, 143–152, 157, 171, 174, 177, 181, 329
    anchoring, 59, 68, 145, 148
    backfill, 46, 65–69, 101–103, 144–145, 148–149, 151, 167, 172, 178, 181

[Underground storage system, Tank]
bedding, 145, 149
leak detection, 10–11, 96, 98, 116, 122–123, 139, 160, 177–179, 333
automatic tank gauging, 50, 61–62, 124, 178, 181
groundwater monitoring, 124, 178–179
interstitial monitoring, 47, 107, 110, 139, 178–179
inventory control, 50, 124, 178
manual tank gauging, 10, 124, 178
statistical inventory reconciliation, 10, 124, 178–179
third-party certification, 178
tightness testing, 10, 58, 68, 124, 140, 146, 178
vacuum, 10, 12, 38, 118–119, 140
vapor monitoring, 10, 124, 178–179
regulatory agencies, 125–135
submersible turbine pump, 10, 60, 62
tank, 2, 4–14, 17, 33, 38, 41–42, 45–46, 50, 52–53, 57–68, 70, 91–93, 95–96, 98–109, 114–119, 121–122, 124–125, 137–141, 143–155, 157, 159, 162, 166–172, 175–181, 183, 186–188, 199–200, 202, 209, 220–222, 227, 229, 230, 233–234, 257, 295, 308, 318, 320, 323–324, 327–330, 332
bulkheads, 43
bushing, 58, 109, 115, 119, 139, 146, 150, 328
compartment, 17, 43, 95, 117, 138–140
[Underground storage system, Tank]
double wall, 38, 42–45, 53, 57–58, 95, 107–108, 117–119, 138–139, 146, 157, 162, 178
hold–down straps, 58, 148
monitoring port, 9, 43–44, 146
observation wells, 62
overfill prevention, 9–10, 96, 116, 124, 180, 333
primary, 42, 44–46, 59, 117–119, 140–141, 146, 151, 162
secondary containment, 8–10, 42, 46, 95, 107, 115, 117, 119, 124, 140, 141
single wall, 38, 41–43, 95, 102, 108, 117–119, 138–139, 146, 157, 159
spill prevention, 10, 116, 124, 179
Underwriters Laboratories (UL), 2–3, 5, 7, 11–12, 14, 16–17, 33, 40, 43, 45, 50–51, 56–58, 61, 63, 72, 75–76, 79, 81, 92, 97, 99–111, 113, 121, 137–139, 141, 160, 177, 184, 187–189, 191, 193–198, 208, 210, 212, 214, 232–233, 256, 260, 262, 274, 277–282, 285, 295, 309, 322, 328, 331–332, 338
code compliance verification list, 189, 196, 197, 309
Industry Advisory Conference, 103, 105, 189
Underwriters' Laboratories of Canada (ULC), 51, 113–120, 189, 199–208, 328, 331, 332–333, 338
Uniform Fire Code (UFC), 8–9, 14–16, 26, 29, 31–32, 55, 73, 75–78, 80–81, 83, 188, 191, 196, 214, 226–227, 233–234, 256, 268, 278, 327, 330, 332

U.S. Coast Guard spill data, 246, 267
U.S. Department of Commerce, 248
U.S. Department of Energy, 152
U.S. Department of Housing and Urban Development, 228–232, 234
U.S. Department of Labor:
  Occupational Safety and Health Administration, 56, 73, 76, 255, 327, 330, 338
U.S. Department of Transportation (DOT), 9, 67–68, 86–87
U.S. Environmental Protection Agency (EPA), 5–6, 11–14, 46, 54, 56, 58, 71, 95, 122, 153–155, 157, 159, 162, 178, 180, 185, 217–222, 245, 252, 256, 327, 330, 335–337
  chemical-resistant liner requirements, 252
  Clean Water Act, 14, 73, 217, 327
  Facility Response Plan, 222
  50-state study, 10, 154–155
  40 CFR 280, 50–51, 56, 95, 122, 175–177, 308, 327
  Interim Prohibition, 9, 123, 154, 157
  Oil Pollution Act, 73, 217, 245, 330
  Spill Prevention Control and Countermeasures, 15–16, 73, 93, 185, 217–222, 245, 327, 330
Utah State University, 100–101, 104

Vandalism, 15, 92, 93
Vaults, 17, 234
[Vaults]
  aboveground, 183, 200, 255, 295
  underground, 17, 188–189, 233, 255–256, 332
Vent, 49, 117, 180, 197, 225–226, 256, 263, 265, 300, 304, 306, 312, 319–322, 332–333
[Vent]
  discharge capacity, 262–264, 266, 284
  discharge rate, 260–262, 264, 266, 284, 285
  minimum size, 108, 194, 202–203, 257

Watkins, Dr. Reynold, 100–102, 104
Weak-roof-to-shell vent design, 47
Welding, 273–274
  automatic processes, 35
    flux core arc, 36
    gas metal arc, 36
    metal inert gas (*see* gas metal arc)
    shielded metal arc, 36
    submerged arc, 37
  electrode, 34, 36, 37
Weld seam, 38, 43, 146, 309
  butt weld, 34
  lap weld, 34
Western Fire Chiefs Association, 26, 31–32, 338

Zinc, 7, 45, 58, 115, 119, 161, 171, 176, 328
Zoning, 91, 95, 225, 227, 228, 234, 322